Understanding Smart Sensors

Second Edition

For a listing of recent titles in the *Artech House Sensors Library,*
turn to the back of this book.

Understanding Smart Sensors

Second Edition

Randy Frank

Artech House
Boston • London

Library of Congress Cataloging-in-Publication Data
Frank, Randy.
      Understanding smart sensors / Randy Frank.—2nd ed.
          p.   cm.—(Artech House sensors library)
      Includes bibliographical references and index.
      ISBN 0-89006-311-7 (alk. paper)
              1. Detectors—Design and construction.   2. Programmable controllers.
          3. Signal processing—Digital techniques.   4. Semiconductors.
          5. Application specific integrated circuits.   I. Title.
      TA165.F724   2000
      681'.2—dc21
                                                              00-021296
                                                              CIP

British Library Cataloguing in Publication Data
Frank, Randy
      Understanding smart sensors.—2nd ed.—(Artech House sensors library)
      1. Detectors—Design and construction   2. Programmable controllers
      3. Signal processing—Digital techniques   4. Application specific integrated circuits
      I. Title
      681.2
      ISBN 0-89006-311-7

Cover and text design by Darrell Judd
Cover image courtesy of Sandia National Laboratories

International Standard Book Number: 0-89006-311-7
Library of Congress Catalog Card Number: 00-021296

10 9 8 7 6 5 4 3 2 1

Dedicated to the memory of the one person who would have loved to see this book but did not—my father, Carl Robert Frank.

Contents

Preface

The number one challenge facing engineers is rapidly changing technology, according to a 1999 survey [1]. IBM estimates that more data has been generated in the past 30 years than in the previous 5,000 years! There certainly is no lack of available information on what is happening and the ongoing activities in the area of sensing, micromachining, and microelectromechanical systems (MEMS). The difficulty comes in making effective decisions as users of the technology or establishing a long-range plan of where a company might use existing and future products to develop end-user products. This book condenses the existing material into a highly readable format and links the variety of ongoing activities in the smart sensor and MEMS area.

According to Dana Gardner and as noted in the first edition of this book, "By the year 2000, 50% of all engineers will design with sensors, up from 16% who routinely used them at the beginning of the decade" [2]. I do not know if that prediction came true, but the 1990s certainly should be remembered as the decade when MEMS technology accelerated from the laboratory into production and the decade that established smart sensors through the IEEE 1451 standard. Micromachining technology will continue to be the primary reason for sensors achieving cost breakthroughs that allow widespread sensor usage. At the heart of most smart sensors will be digital integrated circuit technology.

Embedded microcontrollers already play a hidden role in most of the common activities that occur in our daily lives. Use a cellular phone, receive a page, watch television, listen to a compact disc, or drive a current model car and you have the assistance of embedded microcontrollers. For example, there are a dozen or so microcontrollers in a typical car, over 50 if you drive a well-equipped luxury vehicle. Semiconductor sensors provide many of the inputs to those devices. The number of sensors and the intelligence level are increasing to keep up with increasing control complexity.

Semiconductor sensors initially were developed to provide easier-to-interface, lower-cost, and more reliable inputs to electronic control systems. The microcontrollers at the heart of these systems have increased in complexity and capability while drastically achieving reduced cost per function. Semiconductor technology has also been applied to the input side for a few sensor inputs (pressure, temperature, acceleration, optoelectronics, and Hall-effect devices) but is just starting to broaden in scope (level of integration) and sensed parameters and to achieve some of the cost-reduction benefits from integration.

The system outputs have done a better job of keeping up with advances in semiconductor technology. The term *smart power* refers to semiconductor power technologies that combine an output power device(s) with control circuitry on the same silicon chip. Both input and output devices are receiving greater focus, the capability of combining technologies is being extended, and the need for systems-level communications is finally making smart sensors a reality.

Wen Ko of Case Western Reserve University established a vision for intelligent sensors [3], but Joe Giachino of Ford Motor Company is frequently given credit for the term *smart sensor*, based on his 1986 paper [4]. Several others, including Middelhoek and Brignell, claim part of the credit for pioneering the concept of smart sensors with capabilities beyond simple signal conditioning. The communication of sensory information is finally requiring consensus for the true meaning of smart sensor.

The ultimate capabilities of new smart sensors will undoubtedly go far beyond today's projections. An understanding of what is possible today and what can be expected in the future is necessary to take the first step toward smarter sensing systems. This book is intended to provide the reader with knowledge regarding a broad spectrum of possibilities based on current industry, university, and national laboratories' R&D efforts in smart sensors. It discusses many recent developments that will affect sensing technology and future products.

In this second edition, every chapter has been reviewed, and new, more current material has been added. The recent balloting and acceptance of IEEE 1451.1 and 1451.2 provided the impetus for updating the first edition. Chapters 12 and 13 address those important additions to the future of smart sensing.

I would like to extend my sincere appreciation to Mark Shaw, whose concept of the phases of integration became an underlying theme for this book and, I believe, the way that smart sensing will evolve. It certainly has held true for well over 6 years. A number of other people played an important role in making this book a reality:

- Ray Weiss of *Computer Design* magazine provided methodology guidance and was the prime mover.
- Mark Walsh and the team at Artech House were very supportive at every step in the process.
- Lj Ristic, Mark Shaw, Cindy Wood, and Mark Reinhard from Motorola Semiconductor Products Sector provided chapter reviews.
- Carl Helmers and the folks at *Sensors* magazine and Sensors Expo provided many publishing opportunities that were helpful in documenting several aspects of smart sensors.
- Sandia National Laboratories provided material for cover artwork.

Finally, this book would not have been possible without the critical evaluation, tolerance, and encouragement of my wife, Rose Ann. Any expression of appreciation is small compared to the sacrifices she made.

References

[1] *Design News,* Jan. 18, 1999, p. 74.

[2] Gardner, D. L., "Accelerometers for Exotic Designs," *Design News,* July 17, 1989, p. 55.

[3] Ko, W. H., and C. D. Fung, "VLSI and Intelligent Transducers," *Sensors and Actuators,* 2 (1982), pp. 239–250.

[4] Giachino, J. M., "Smart Sensors," *Sensors and Actuators,* 10 (1986), pp. 239–248.

1

Smart Sensor Basics

A rose by any other name would smell as sweet.
—William Shakespeare

A rose with a microcontroller would be a smart rose.
—Randy Frank

1.1 Introduction

Just about everything today in the technology area is a candidate for having the prefix *smart* added to it. The term *smart sensor* was coined in the mid-1980s, and since then several devices have been called smart sensors. The intelligence required by such devices is available from microcontroller unit (MCU), digital signal processor (DSP), and application-specific integrated circuit (ASIC) technologies developed by several semiconductor manufacturers. Some of those same semiconductor manufacturers are actively working on smarter silicon devices for the input and output sides of the control system as well. The term *microelectromechanical system* (MEMS) is used to describe a structure created with semiconductor manufacturing processes for sensors and actuators. To understand what is occurring today when advanced microelectronic technology is applied to sensors, a brief review of the transitions that have occurred is in order.

Before the availability of microelectronics, the sensors or transducers used to measure physical quantities, such as temperature, pressure, and flow, usually were coupled directly to a readout device, typically a meter that was read by an observer. The transducer converted the physical quantity being measured to a displacement. The observer initiated system corrections to change the reading closer to a desired value. The typical blocks of a measurement system are shown in Figure 1.1 [1].

Many home thermostats, tire pressure gauges, and factory flow meters still operate in the same manner. However, the advent of microprocessor technology initiated the requirement for sensors to have an electrical output that could be more readily interfaced to provide unattended measurement and control. That also required the analog signal level to be amplified and converted to digital format prior to being supplied to the process controller. Today's MCUs and analog-to-digital (A/D) converters typically have a 5V power supply, which has dictated the supply voltage for many amplified and signal conditioned sensors. However, the reduction in the supply voltage from 5V to 3.3V and even lower voltages and the presence of more than one voltage in a system pose challenges not typically associated with even the smartest sensors. Separate integrated circuits (ICs) are available to handle the variety of voltages and resolve the problem, but they add to system and sensor complexity.

Commonly used definitions for the terms *sensor* and *transducer* must be the first in the list of many terms that will be defined. A transducer is a device that converts energy from one domain into another, calibrated to minimize the errors in the conversion process [2]. A sensor is a device that provides a useful output to a specified measurand. The sensor is a basic element of a transducer, but it also may refer to a detection of voltage or current in the electrical regime that does not require conversion. Throughout this book, the terms are used synonymously, because energy conversion is part of every device that is discussed. The mechanical measurements that require a transducer to provide an electrical output are listed in Table 1.1.

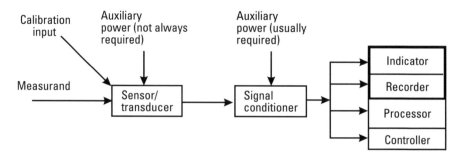

Figure 1.1 General sensing system.

Table 1.1
Mechanical Measurements

Measurement	Typical/Common Techniques
Displacement/position	Variable reluctance, Hall effect, optoelectronic
Temperature	Thermistor, transistor base-emitter voltage (Vbe)
Pressure	Piezoresistive, capacitive
Velocity (linear/angular)	Variable reluctance, Hall effect, optoelectronic
Acceleration	Piezoresistive, capacitive, piezoelectric
Force	Piezoresistive
Torque	Optoelectronic
Mechanical impedance	Piezoresistive
Strain	Piezoresistive
Flow	Δ pressure, or delta pressure
Humidity	Resistive, capacitive
Proximity	Ultrasonic
Range	Radar
Liquid level	Ultrasonic
Slip	Dual torque
Imminent collision	Radar

The definition of *smart sensor* (intelligent transducer) has not been as widely accepted and is subject to misuse. However, an Institute of Electrical and Electronics Engineers (IEEE) committee has been actively consolidating terminology that applies to microelectronic sensors. The recently approved IEEE 1451.2 specification defines a smart sensor as a sensor "that provides functions beyond those necessary for generating a correct representation of a sensed or controlled quantity. This function typically simplifies the integration of the transducer into applications in a networked environment" [2]. That definition provides a starting point for the minimum content of a smart sensor. Future smart sensors will be capable of much more, and additional classifications (e.g., smart sensor type 1) may be required to differentiate the products. Chapter 12 addresses the IEEE-approved smart transducer interface for sensors and actuators that establishes a standard for transducer-to-microprocessor communication protocols in a transducer electronic data sheet (TEDS) format. That standard and others will provide further definition and differentiation for smart sensors and accelerate development and commercialization of smart sensors.

1.2 Mechanical-Electronic Transitions in Sensing

An early indication of the transition from strictly mechanical sensing to electronic techniques is demonstrated in the area of temperature and position measurements. Thermistors and semiconductor temperature sensors that were lower in cost, smaller in size, and easier to interface to other circuit elements replaced thermocouples and expansion thermometers. In position measurements, variable-reluctance sensors with magnetic pickups have been replaced by Hall-effect, optical, and magnetoresistive (MR) sensing elements. All those techniques make use of a previous problem that detracted from the ideal performance of a transistor or an IC. The sensitivity of transistors to temperature, light, magnetic fields, stress, and other physical variables is exploited in many existing semiconductor sensors.

The expanding range of parameters that can be sensed using semiconductor technology is part of the increasing interest in smart sensing. With micromachining, or chemical etching techniques, mechanical structures have been produced in silicon that have greatly expanded the number and types of measurements that can be made. For example, a rubber diaphragm connected to a potentiometer can be replaced by a silicon diaphragm and piezoresistive elements for measuring pressure. That approach has been used in production sensors for over two decades. More recently, cantilever beams and other

Table 1.2
Sensing Techniques

Technique	Status in Silicon Sensor
Piezoresistive	Pressure, acceleration
Capacitive	Pressure, acceleration, position
Piezoelectric	Pressure, acceleration, vibration
Optoelectronic	Position, velocity
Magnetic	Position, velocity, magnetic field
Radar	Limited production
Laser infrared radar (lidar)	Production/research/development
Ultrasonic	Production

suspended structures have been manufactured in silicon, and acceleration is measured by resistive, capacitive, or other techniques. Table 1.2 lists a number of sensing techniques and their status relative to implementation in silicon sensors. (Chapter 2 explains the most popular developments in micromachining that allow sensors and other MEMS devices to be fabricated.)

Sensor manufacturers are adapting processes used to manufacture advanced semiconductor technologies. As a result, sensors are being manufactured, either concurrently or separately, that take advantage of the performance enhancements that IC technology can provide, and a significant step forward is occurring in sensing technology. (Chapters 3 and 4 develop those interfacing and integration aspects.)

1.3 Nature of Sensors

The output from most sensing elements is low level and subject to several signal interference sources, as shown in Figure 1.2, a generalized model of a transducer [3]. Self-generating transducers such as piezoelectric devices do not require a secondary input to produce an output signal. However, transducers based on resistive, capacitive, and inductive sensing elements require excitation

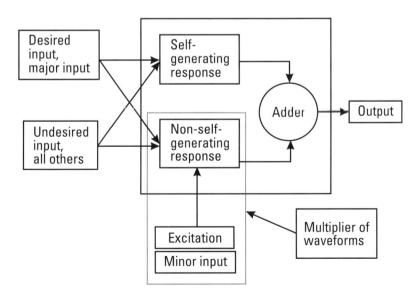

Figure 1.2 General transducer model. (*After:* [3].)

to provide an output. In addition to the desired input (e.g., pressure and unde-sired environmental effects, such as temperature, humidity, and vibration) are factors that affect the performance and accuracy of the transducer, factors that must be taken into account during the design of the transducer. Compensation for those secondary parameters historically has been performed by additional circuitry, but with smart sensing technology the compensation can be inte-grated on the sensor or accomplished in the microcontroller.

The output of a micromachined piezoresistive silicon pressure sensor and the effect of temperature on both the span and the offset are demonstrated in Figure 1.3 [4]. Although the output is quite linear, in this case within 0.1% full scale (F.S.), the output varies due to the effect of temperature on the span of the sensor by about 0.12 mV/°C. Because that signal level is insufficient to directly interface to a control IC, additional amplification and calibration typically are performed in the next stage of a transducer.

In a simple control system, the sensor is only one of three items required to implement a control strategy. The sensor provides an input to a controller with the desired strategy in its memory, and the controller drives an output stage to modify or maintain the status of a load, such as a light, a motor, a sole-noid, or a display. As shown in Figure 1.4, a signal conditioning interface typi-cally exists between the sensor(s) and the controller and between the controller

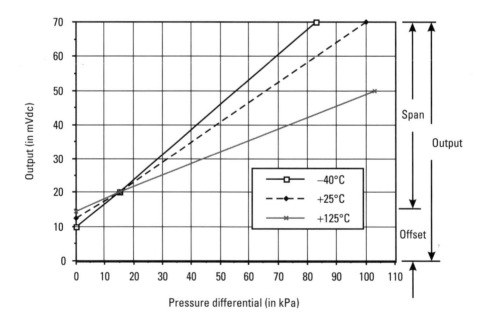

Figure 1.3 Effect of temperature on piezoresistive pressure sensor output.

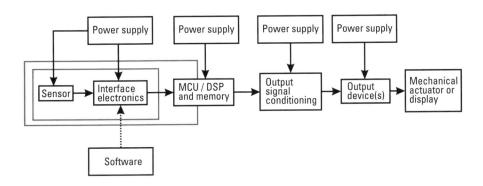

Figure 1.4 Generic control system.

and the output device. Smart sensing includes a portion of the controller's functions in the sensor portion of the system. That means software will play an increasingly important role in smart sensors. The power supply requirements for the electronics and the sensor represent an additional consideration that is becoming more important as MCU voltages are decreased and more sensors are used in battery power or portable applications. The number of supplies in Figure 1.4 may not be required for a particular application, but they serve as a reminder for considering the available voltage for the sensor and the interface versus the rest of the system.

The smart sensor models developed by several sources [5–7] have as many as six distinct elements for analog sensors. As shown in Figure 1.5, in addition to the sensing element and its associated amplification and signal conditioning, an A/D converter, memory of some type, and logic (control) capability are

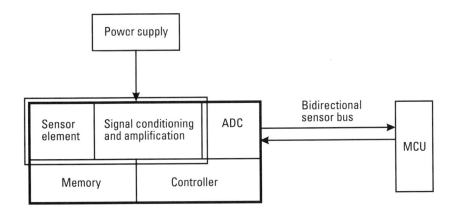

Figure 1.5 Smart sensor model.

included in the smart sensor. Once the signal is in digital format, it can be communicated by several communication protocols. The regulated power supply also required for the system and its effect on system accuracy must be taken into account. That is becoming more of an issue as power management issues are addressed in system design and different supply voltages proliferate.

Reducing the number of discrete elements in a smart sensor (or any system) is desirable to reduce the number of components, form factor, interconnections, assembly cost, and frequently component cost as well. The choices for how that integration occurs are often a function of the original expertise of the integrator. For example, as shown in Figure 1.6, a sensor manufacturer that already uses semiconductor, that is, bipolar or metal oxide semiconductor (MOS), technology for the sensing element may expand the capability and increase the value (and intelligence) of the sensing unit it produces by combining the signal conditioning in the same package or in a sensor module. Through integration, the signal conditioning can also be combined at the same time the sensor is fabricated or manufactured.

While the process of integration is more complex, the integrated sensor can be manufactured with the sensor and signal conditioning optimized for a particular application. Conversely, an MCU manufacturer using a complementary metal oxide semiconductor (CMOS) process typically integrates memory, A/D, and additional signal conditioning to reduce the number of components in the system. A variety of combinations are indicated in Figure 1.6. Processing technology is a key factor. However, manufacturers not only must be willing to integrate additional system components, they also must achieve a cost-effective solution. Combinations of hybrid (package level) and monolithic integration are discussed frequently in the remainder of this book. Different design philosophies and the necessity to partition the sensor/system at different points can determine whether a smart sensor is purchased or, alternatively, designed using a sensor signal processor or other components necessary to meet the desired performance of the end product.

The integration path can have a significant effect on the ultimate level of component reduction. As shown in Figure 1.7 [8], the input (demonstrated by a pressure sensor), computing (high-density CMOS [HCMOS] microcontroller), and output side (power MOS) are all increasing the level of monolithic integration. The choice of sensor technology, such as bipolar, can have a limiting effect on how far the integration can progress. For example, a bipolar sensor can increase integration level by adding signal conditioning and progress to a monolithic level III sensor. Through package-level integration, a two-chip sensor controller can be achieved by combining the sensor with an HCMOS

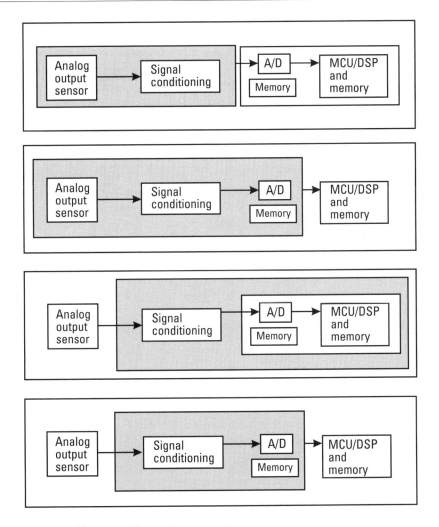

Figure 1.6 Partitioning and integration possibilities.

microcontroller. However, the highest level of monolithic integration, level V, will be realized only by pursuing MOS-compatible sensing and power-control technologies.

Realizing the full potential of those new sensors will require a new approach to identify sensor applications. The list of sensor terms in Table 1.3 serves as a starting point for rethinking the possibilities for smart sensors. Many of the terms are associated with a system and not the sensor portion of the system.

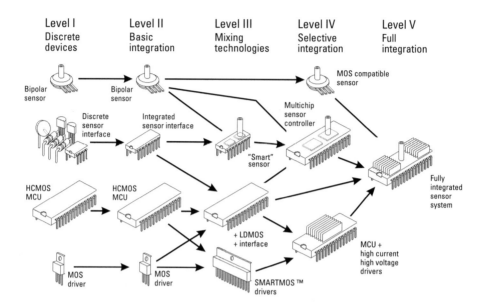

Figure 1.7 Sensor technology migration path.

Table 1.3
Terms Used to Identify New Smart-Sensor Applications

Measure	Understand
Monitor	Diagnose
Correct	Control
Detect	Presence/absence
Safe operating area	Exceed level
Communicate	Identify
Prevent failure	Maintain
Warning	Instrument
Regulate	Gauge
Observe	Know
Determine	See/hear/touch/smell/taste
	How/what/where/when/why

When sensors are combined with an MCU, a DSP, or an ASIC, the capability to obtain additional performance improvements is limited only by the capability of the computing element and the imagination of the designer. Chapters 5, 6, and 7 explain and develop some of those possibilities. In addition to signal transmission in distributed control systems through a variety of protocols described in Chapter 6, the possibilities of portable, wireless, and remote sensing are explored in Chapter 8. A broad variety of micromechanical elements and additional system components are investigated in Chapter 9. Among the challenges and limiting factors to higher levels of integration are packaging, testing, and reliability. Chapter 10 reveals the progress being made in these areas for production sensors. However, packaging, testing, and reliability are some of the formidable challenges that must be addressed for smart sensors. The combination of the previously discussed aspects is already being researched, as discussed in Chapter 11, with the addition of the system output and sensing systems.

Chapter 12 presents what could be the milestone in the future for the turning point in the evolution of smart sensors. The availability and acceptance of standards can accelerate the development and use of smart sensors. Chapter 13 provides some examples and explores the implications of standards that affect sensing. Finally, based on the system-level complexity already possible and continuously evolving, a look into the not-too-distant future of smart sensing is shared in Chapter 14.

1.4 Integration of Micromachining and Microelectronics

Increasing the performance and reliability and reducing the cost of electronic circuits through increased integration are standard expectations for semiconductor technology. In the area of semiconductor sensors, however, that integration has been limited to Hall-effect and optoelectronic devices. The recent combination of micromechanical structures, sensing elements, and signal conditioning is the beginning of a new chapter in sensor technology. The combination of microelectronics with micromechanical structures promises to change future control systems and enable entirely new applications that previously were too costly for commercial purposes. The term *microelectromechanical systems*, or sometimes simply *microsystems*, is used to describe the structures and functions provided by micromachining and the addition of microelectronics to those structures.

A sensor with its own dedicated interface circuitry has several advantages. The sensor designer can trade off unnecessary performance characteristics for those that will provide desirable performance advantages to the sensor-interface combination. Normally, interface ICs are designed for a broad range of applications and such tradeoffs are not possible. The combination allows the sensor user to treat the sensor as a "black box" and easily design a complex control system.

The integrated sensor takes advantage of integrated temperature sensing to more closely track the temperature of the sensing element and compensates for the effects of temperature over the temperature range. By reducing the number of internal connections, the reliability of integrated sensors is inherently better than a separate sensor and control circuit, even when the separate components are manufactured using a thin-film ceramic substrate. For a four-terminal sensor element, a reduction from 23 to only 9 connections is possible using an integrated solution. Because the sensor provides the first information to a control system, the reliability of that information is critical to the entire system's reliability.

Figure 1.8 is useful in analyzing the results of integrating the first two stages illustrated in Figure 1.7 to obtain the sensor portion of a level III system. Figure 1.8 is an enlarged view of a fully integrated piezoresistive pressure sensor with the amplification and signal conditioning circuitry, including laser-trimmed resistors, located around the silicon diaphragm etched in the middle of the chip. The single piezoresistive element is one arm of the X that is near the

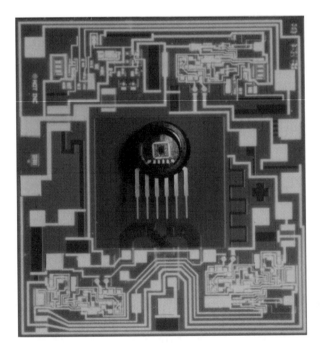

Figure 1.8 Photomicrograph of pressure sensor die with packaged sensor over the micromachined diaphragm. (Courtesy of Motorola, Inc.)

edge of the right side of the diaphragm. The package used for that sensor (shown in the middle of the diaphragm with the actual chip inside) has six terminals. Only three terminals are required for the application; the other three are used to access measurement points on the chip during the trim process and are not necessary for the sensor to function properly.

A major advantage of integrated sensing and signal conditioning is the addition of calibration through on-chip techniques, such as laser trimming of thin-film resistors on the sensor die, and the subsequent ability to obtain part-for-part replaceability at the component level. The amount of signal conditioning can vary. For example, the addition of thin-film resistors and laser trimming to the sensing element is all that is necessary to produce a calibrated and temperature-compensated sensor for disposable blood pressure applications. For lower volume applications, the requirement for unique transfer function or interface circuitry may not be cost effective. In those cases, the basic sensor with external circuitry is still the best choice. Eventually, the unique lower volume applications could also benefit from the advantages of integration.

Sensing and integrated sensing (sensing plus signal conditioning) are somewhat analogous to other mixed-signal processes that exist in semiconductors, especially in power and smart-power technology, that is, the output side of the control system. Today's smart-power technologies integrate bipolar and CMOS circuitry with multiple power metal oxide semiconductor field effect transistor (MOSFET) output devices. The process is more complex than a discrete power MOSFET, but the performance achieved by the combination of technologies provides a specific function, component reduction for increased reliability, and space reduction for lower cost assemblies and more than justifies the higher processing cost. After a number of years of process and design improvements, smart-power devices have established broad market acceptance, especially for custom designs. Similarly, smarter sensors that have their own signal conditioning circuitry onboard are in the early phases of market acceptance and are approximately at the same stage of development as smart power was five to seven years ago. However, smart sensors may not have to be custom devices to satisfy a large number of applications.

New packages must be developed for smart sensors to accommodate the additional connections for power, ground, output, diagnostics, and other features that the combination of technologies can provide. Sensors have few, if any, commonly accepted packages. Instead, each supplier provides a unique pinout and form factor. The problem is exacerbated by the addition of more features to sensors, either through integration or by the addition of circuitry.

Prior to the era of sensor integration, products that combine technology at the package level, rather than at the silicon level, have been the industry norm. A hybrid or module solution has the advantage of using proven available technology to achieve a more sophisticated product solution. That may be in a printed circuit board or ceramic substrate form. Figure 1.9 shows the steps that can be taken toward a higher level of monolithic integration.

Figure 1.10 is an example of a level IV, two-chip smart sensor. The unit uses the integrated pressure sensor showed in Figure 1.8, along with a single-chip 8-bit MCU with an onboard A/D converter and an electrically erasable programmable read only memory (EEPROM) to achieve a minimum-component-count smart sensor [9]. Except for three resistors used for increased resolution, the other 10 components in the circuit are necessary for proper functioning of the MCU. An undervoltage-sensing circuit is used for the reset function to provide an orderly powerdown in the case of low battery voltage.

Two jumpers (and *no* potentiometers) are used for the initial calibration. The 01 code for the jumpers (J1 and J2) is used for zero calibration, and a 10 code is used to indicate full-scale value. A 000 on the display indicates zero, and FFF indicates the full-scale value during calibration. Either the 00 or 11 code is used for normal operation. The values of zero and full scale are stored in the EEPROM. Three resistors are used to provide a ratiometric reference of

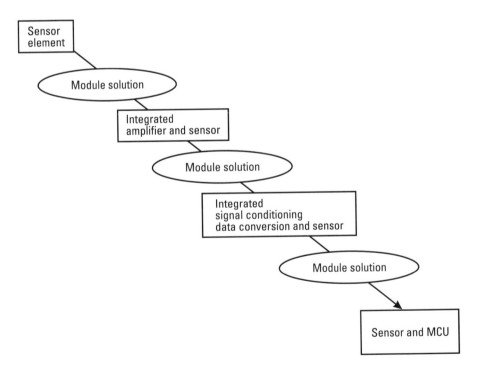

Figure 1.9 Steps of development toward increased integration.

Figure 1.10 Two-chip smart sensor for pressure measurements. (Courtesy of Motorola, Inc.)

V_{RL} = 0.3 to V_{RH} = 4.7V, utilizing the highest output capability of the integrated sensor.

The liquid crystal display (LCD) is driven directly from the MCU by using the MCU to generate a square wave signal every 32 ms. The backplane is alternately inverted with the segments and out-of-phase with the segment that is on. Each MCU port is connected to a digit, and two additional input-output (I/O) lines are used to generate the decimal point and the backplane signal. That approach utilizes software to eliminate the need for an additional display driver and achieves a reduction in both component count and space. The memory required by the MCU is minimal, only 1 KB read only memory (ROM) and 4 bytes EEPROM, leaving essentially all the remaining memory for other functions. The program for the MCU is written in C, a high-level language.

1.5 Summary

Today, it is possible to build a smart sensor for several measurements with basically two semiconductor components: the sensor and the MCU. That is the current phase of development and one of the necessary steps to the next level. The smart money is being placed on areas that take advantage of the number of technologies that are available or in developments that will fundamentally change the nature of sensing, control systems, and aspects of everyday life. The remaining chapters discuss a variety of aspects of sensing and state-of-the-art developments that will allow those who understand how to apply such developments to create the next-generation products and systems.

References

[1]   Beckwith, T. G., N. L. Buck, and R. D. Marangoni, *Mechanical Measurements*, Reading, MA: Addison-Wesley, 1982.

[2]   IEEE 1451.2 Standard, "A Smart Transducer Interface for Sensors and Actuators—Transducer to Microprocessor Communication Protocols and Transducer Electronic Data Sheet (TEDS) Formats," Piscataway, NJ: IEEE Standards Department, 1998.

[3]   Wright, C., "Information Conversion Separates Noise Levels So You Can Control Them," *Personal Engineering & Instrumentation News*, Apr. 1993, pp. 63–67.

[4]   *Sensor Device Data/Handbook*, DL200/D Rev. 4, 1998, Motorola Semiconductor Products Sector, Austin, TX.

[5]   Najafi, K., "Smart Sensors," *J. Micromechanics and Microengineering*, Vol. 1, 1991, pp. 86–102.

[6]   Ina, O., "Recent Intelligent Sensor Technology in Japan," *Soc. Automotive Engineers*, SAE891709, 1989.

[7]   Maitan, J., "Overview of the Emerging Control and Communication Algorithms Suitable for Embedding Into Smart Sensors," *Proc. Sensors Expo*, Cleveland, Sept. 20–22, 1994, pp. 485–500.

[8]   Benson, M., et al., "Advanced Semiconductor Technologies for Integrated Smart Sensors," *Proc. Sensors Expo*, Philadelphia, Oct. 26–28, 1993, pp. 133–143.

[9]   Frank, R., "Two-Chip Approach to Smart Sensors," *Proc. Sensors Expo*, Chicago, 1990, pp. 104C1–104C8.

Select Bibliography

Ko, W. H., and C. D. Fung, "VLSI and Intelligent Transducers," *Sensors and Actuators*, 2 (1982), pp. 239–250.

Microelectromechanical Systems Advanced Materials and Fabrication Materials, Washington, D.C.: National Academy Press, 1997.

Middelhook, S., and A. C Hoogerwerf, "Smart Sensors: When and Where?" *Sensors and Actuators*, 8 (1985), pp. 39–48.

Muller, R. S., et al. (eds.), *Microsensors*, IEEE Press, 1991.

Trimmer, W., "Micromechanics and MEMS," *Classic and Seminal Papers to 1990*, IEEE Press, 1997.

2

Micromachining

I wonder how many angels can dance on the head of a pin?
—Unknown Middle Ages philosopher

2.1 Introduction

The answer to that question is no more than 25,200, given that a pin has a diameter of $\frac{1}{16}$ inch (15,875 μm) and each (micromachined) angel has a diameter of 100 μm. So much for philosophy. Micromachining is causing the reinvestigation of every aspect of physics, chemistry, biology, and engineering. Thermodynamics, mechanics, optics, fluidics, acoustics, magnetics, electromagnetics, wave, kinetics, and nuclear forces, as well as medicine, robotics, displays, and instrumentation, are being investigated in academic, national, and industrial research and development (R&D) labs. Micromachining technology is enabling the extension of semiconductor-based sensing beyond temperature, magnetic, and optical effects to produce mechanical structures in silicon and sense-mechanical phenomena. Many universities are creating cross-departments and cross-college programs to explore the future possibilities of micromachined MEMS devices.

Micromachining is, in the most common usage, a chemical etching process for manufacturing three-dimensional microstructures that is consistent with semiconductor processing techniques. IC manufacturing processes used to make the microstructures include photolithography, thin-film deposition, and chemical and plasma etching. Bulk micromachining has been used to manufacture semiconductor pressure sensors since the late 1970s. Recently, newer

17

techniques such as surface micromachining have been developed that achieve even smaller structures. In addition, the processing techniques for surface micromachining are more compatible with the CMOS processes used to manufacture integrated circuits.

Silicon has many properties that make it ideal for mechanical structures. As indicated in Table 2.1, it has a modulus of elasticity (Young's modulus) comparable to steel and a higher yield strength than steel or aluminum [1, 2]. Silicon has essentially perfect elasticity, resulting in minimal mechanical hysteresis. (It is, however, a brittle material and will crack when stressed beyond its elastic limits.) Also, silicon's electrical properties have made it the material of choice in most integrated circuits, providing established manufacturing techniques for many aspects of micromachined sensors. Micromachined semiconductor sensors take advantage of both the mechanical and electrical properties of silicon, but products that fully exploit the combination of the mechanical and electrical properties are still in their infancy.

Table 2.1
Properties of Silicon Compared to Other Materials

Property	3C-SiC (6H-SiC)	GaAs	Si	Diamond
Melting point (°C)	Sublimes at 1,825	1,238	1,415	Phase change
Max. operating temp. (°C)	873 (1,240)	460	300	1,100
Thermal conductivity (W/cm °C)	4.9	0.5	1.5	20
Thermal expansion coeff. ($*10E-6 °C^{-1}$)	3.8 (4.2)	6.9	2.6	—
Young's modulus (GPa)	448	75	190	1,035
Physical stability	Excellent	Fair	Good	Fair
Energy gap (eV)	2.2 (2.9)	1.42	1.12	5.5
Electron mobility (cm^2/V s)	1,000 (500)	8,500	1,350	2,200
Hole mobility (cm^2/V s)	40 (50)	400	600	1,600
Sat. electron drift vel. ($*10E7$ cm/s)	2.5 (2)	2	1	2.7
Breakdown voltage ($*10E7$ cm/s)	3 (4-6)	0 .4	0.3	10
Dielectric constant	9.7	13.2	11.9	5.5
Lattice constant (Å)	4.36	5.65	5.43	—-

The relative ease of accomplishing both bulk and surface micromachining has led to many researchers investigating a variety of applications. Some of the areas being investigated will lead to smarter sensors through higher levels of integration. A key process associated with micromachining is the bonding of silicon to a silicon or substrate material. The processes used for micromachining, associated processes, and application of the technology to sensors are covered in this chapter.

2.2 Bulk Micromachining

Bulk micromachining is a process for making three-dimensional microstructures in which a masked silicon wafer is etched in an orientation-dependent etching solution [3]. Using micromachining technology, several wafers can be fabricated simultaneously and lot-to-lot consistency is maintained by controlling a minimal number of parameters. Key parameters in bulk micromachining include crystallographic orientation, etchant, etchant concentration, semiconductor starting material, temperature, and time. Photolithography techniques common in IC technology precisely define patterns for etching both sides of silicon wafers. The crystallographic orientation, etchant, and semiconductor starting material are chosen by design, leaving etchant concentration, temperature, and time as lot-to-lot control items.

Silicon ICs are typically fabricated (manufactured) using <100> or <111> silicon. In bulk micromachining, an anisotropic (unidirectional) etchant, such as ethylene-diamine-pyrocatechol (EDP), hydrazine (N_2H_4), tetramethylammonium hydroxide (TMAH), or potassium hydroxide (KOH), attacks the <100> plane of silicon. The <100> plane is etched at a much faster rate than the <111> plane, typically 35 times faster. N-type silicon is etched at a much faster rate (>50 times faster) than p+-type, so n-type material is often used as the starting material. P+-type material can be epitaxially grown on the wafer or diffused into the wafer to add a further control element in defining the dimensions. Agitation maintains uniform concentration during anisotropic etching. The characteristic shape (preferential etching) of anisotropic etching of <100> silicon is shown in the cross-section of Figure 2.1(a), which produces a 54.7-degree angle for the <100> silicon [4, 5]. The top view of etching into the surface of the silicon appears as a pyramid-shaped pit.

Etch rates of 1.0–1.5 μm/minute occur in the <100> plane of silicon with etch temperatures of 85–115°C for common etchants such as EDP and KOH [6]. Isotropic etching, shown in Figure 2.1(b), has etch rates independent of the crystallographic orientation. Isotropic etching allows undercut and cantilever structures to be produced. In bulk silicon, however, it is more difficult to

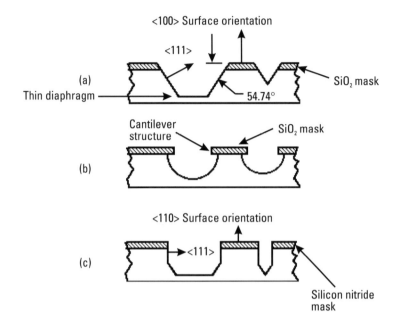

Figure 2.1 Bulk micromachined structures: (a) anisotropic, (b) isotropic with agitation, and (c) alternative crystal orientation and mask material.

control than anisotropic etching, and the level of agitation influences the results. Undercutting and suspended structures are achieved with anisotropic etching with mask patterns and extended etch times. To expand the applications of micromachining, other chemicals, such as sodium hydroxide (NaOH), are used for etching the <100> and <110> plane. Also, other materials such as silicon nitride can be used for the masking layer, as shown in Figure 2.1(c). Table 2.2 lists the etch rates of common materials, and possible etchants are described in Section 2.5.2 [2]. The etch rates for 317 combinations of 16 materials (single-crystal silicon, doped and undoped polysilicon, several types of silicon dioxide, stoichiometric and silicon-rich silicon nitride, aluminum, tungsten, titanium, titanium-tungsten alloy, and two brands of photoresist) used in the fabrication of MEMS devices and ICs in 28 wet, plasma, and plasmaless-gas-phase etches have been reported in one study [7].

Etch stop techniques enhance the accuracy of wet chemical etching. The most common techniques for etch-depth control in bulk micromachining are shown in Figure 2.2 [6]: Precisely controlled diffusions (a), epitaxially grown layers in the silicon crystal (b), or field enhanced depletion layers (c) slow down the etching process at the interface, allowing accurate structures to be obtained. Boron etch stops using EDP or TMAH can be used to produce layers as thick

Table 2.2
Attributes of Bulk Micromachining Etchants (*After:* [2])

Etched Material	Etch Rate (nm/min)	Possible Etchant	Comments
<100> silicon	1,250	EDP	Anisotropic
<110> silicon	1,400	KOH	Anisotropic
Si	900–1,300	SF_6	Isotropic
Si	10,000–300,000	HF	Isotropic
SiO_2	400–8,600	HF	Surface

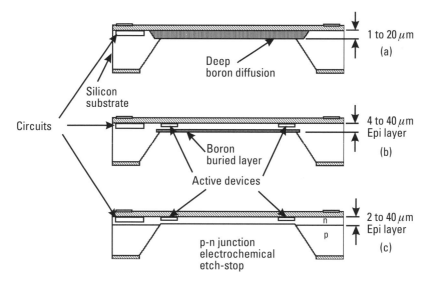

Figure 2.2 Etch-stop techniques in anisotropic etchants: (a) deep boron diffusion, (b) boron buried layer, and (c) p-n junction.

as 15–50 μm, with oxide masking capable of protecting other areas of the chip for adding circuitry.

2.3 Wafer Bonding

In addition to micromachining, different types of wafer bonding are needed to produce more complex sensing structures. The attachment of silicon to a second silicon wafer or silicon to glass is an important aspect of semiconductor sensors. In fact, a clear dependence on wafer bonding as the enabling technology for high-volume MEMS has been identified [8]. Four different approaches to wafer bonding are discussed in this chapter.

2.3.1 Silicon-on-Silicon Bonding

A common approach for manufacturing semiconductor pressure sensors uses a bulk micromachined diaphragm anisotropically etched into a silicon wafer. Piezoresistive sensing elements diffused or ion implanted into the thin diaphragm are either a four-element Wheatstone bridge or a single element positioned to maximize the sensitivity to shear stress [9]. Two silicon wafers are often used to produce the piezoresistive silicon pressure sensor. Figure 2.3 shows a two-layer silicon-on-silicon pressure sensor [9]. The top wafer is etched until a thin square diaphragm approximately 0.001 inch (25.6 μm) in thickness is achieved. The square area and the 54.7-degree angle of the cavity wall are extremely reproducible. In addition to a sealed reference cavity for absolute pressure measurements, the two-layer silicon sensor allows atmospheric or a reference pressure to be applied to one side of the sensor by an inlet hole micromachined in the silicon bottom (constraint) wafer. Several methods are used to attach the top wafer to the bottom, including anodic bonding, glass frit seal, and direct wafer (silicon-to-silicon) bonding or silicon fusion bonding.

The sensor shown in Figure 2.3 uses a glass frit or paste to attach the top wafer to the bottom wafer. The paste is applied to the bottom (constraint) wafer, which is then thermocompression-bonded to the top wafer containing the bulk micromachined pressure sensing structure. The bottom wafer, containing the glass, provides stress isolation and allows a reference vacuum to be sealed inside the combined structure.

To extend the pressure capability to very low-pressure readings ($\leq$2 inches, or 5 cm, of water) and minimize nonlinearity, several different approaches are being pursued. Those approaches include silicon bosses used as stress concentrators in circular, square, and dual rectangular diaphragms; convoluted square diaphragm; and etch stop techniques to control the diaphragm thickness. Diaphragms as thin as 2.5 μm have been used to produce capacitive pressure sensors for 300 mtorr and lower pressure applications [10]. Micromachining can be enhanced by using the electronics capability inherent from semiconductor manufacturing. That may provide an additional solution for low-pressure measurements.

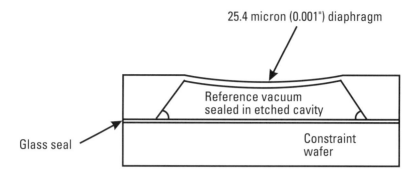

25.4 micron (0.001") diaphragm

Reference vacuum
sealed in etched cavity

Glass seal

Constraint
wafer

Figure 2.3 Pressure sensor with silicon-to-silicon bonding.

2.3.2 Silicon-on-Glass (Anodic) Bonding

Electrostatic, or anodic, bonding is a process used to attach a silicon top wafer to a glass substrate and also to attach silicon to silicon. Anodic bonding attaches a silicon wafer, either with or without an oxidized layer, to a borosilicate (Pyrex®) glass heated to about 400°C when 500V or more is applied across the structure [4]. An example of a product manufactured with anodically bonded silicon to glass is the silicon capacitive absolute pressure (SCAP) sensing element shown in Figure 2.4 [11].

The micromachined silicon diaphragm with controlled cavity depth is anodically bonded to a Pyrex® glass substrate that is much thicker than the silicon die. Feed-through holes are drilled in the glass to provide a precise connection to the capacitor plates inside the unit. The glass substrate is metallized using thin-film deposition techniques, and photolithography defines the electrode configuration. After attaching the top silicon wafer to the glass substrate, the drilled holes are solder-sealed under vacuum. That provides a capacitive sensing element with an internal vacuum reference and solder bumps for direct mounting to a circuit board or ceramic substrate. The value of the capacitor changes linearly from approximately 32 to 39 pF with applied pressure from 17 to 105 kPa. The capacitive element is 6.7 mm by 6.7 mm and has a low temperature coefficient of capacitance (−30 to 80 ppm/°C), good linearity (≈1.4%), fast response time (≈1 ms), and no exposed bond wires.

Silicon is also bonded to a second silicon wafer using anodic bonding. A thin (≈4 μm) layer of Pyrex® glass is sputtered on one of the layers, and a much lower voltage of approximately 50V is applied [4]. Silicon instead of glass for

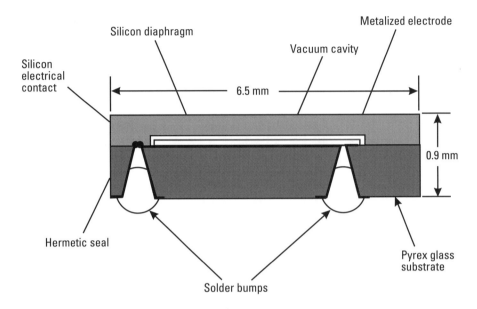

Figure 2.4 SCAP sensor with silicon-on-glass (anodic) bonding.

the support structure allows additional wet-etching techniques to be performed later in the process.

2.3.3 Silicon Fusion Bonding

A technique that bonds wafers at the atomic level without polymer adhesives, a glass layer, or an electric field is known as silicon fusion bonding (SFB) or direct wafer bonding (DWB) [Figure 2.5(a)]. Before bonding, both wafers are treated in a solution, such as boiling nitric acid or sulfuric peroxide [12]. This step covers the surface of both wafers with a few monolayers of reactive hydroxyl molecules. Initial contact of the wafers holds them together through strong surface tension. Subsequent processing at temperatures from 900 to 1,100°C drives off the hydroxyl molecules. The remaining oxygen reacts with the silicon to form silicon dioxide and fuses the two surfaces.

Silicon fusion bonding can be used to reduce the size of a micromachined structure. As shown in Figure 2.5(b), the anisotropically etched cavity can be much smaller, yet the diaphragm area is identical for an SFB pressure sensor compared to a conventional pressure sensor. The bottom wafer with the anisotropically etched cavity is silicon fusion bonded to a top wafer. After bonding, the top wafer is etched back to form a thin diaphragm and the bottom wafer is ground and then polished to open the access to the diaphragm. For absolute pressure sensors, such as the units in Figure 2.5(b), a support wafer is attached to the structure containing the diaphragm produced by any of the previously described methods.

Table 2.3 is a comparison of the typical process conditions for silicon-to-silicon bonding, anodic bonding, and silicon fusion bonding.

2.3.4   Wafer Bonding for More Complex Structures and Adding ICs

Wafer bonding is being investigated as a means to integrate other materials and to combine micromachined structures with microelectronics. Researchers at the University of California, Berkeley, have used an epoxy bond for attaching a sapphire wafer with a gallium-nitride (GaN) film to a silicon wafer. The sapphire wafer is bonded faced down to the silicon wafer. Short, high-intensity pulses from a laser scanned back and forth across the sapphire wafer separated the GaN from the sapphire. The GaN that remained attached to the silicon wafer did not have irregularities in the film [13].

Sandia National Laboratories has grown a GaN film on a nonstandard <111> silicon wafer. A hydrogen implant on the <111> wafer causes small bubbles to form when the wafer is attached to a traditional (high-volume) <100> wafer using silicon-to-silicon bonding. When the wafer is heated during bonding, the small bubbles expand, breaking off a thin <111> layer of silicon. The resulting silicon on insulator system has GaN that can also integrate CMOS circuits [13].

2.4   Surface Micromachining

The selective etching of multiple layers of deposited thin films, or surface micromachining, allows movable microstructures to be fabricated on silicon wafers [14]. With surface micromachining (shown in Figure 2.6), layers of structural material, typically polysilicon, and a sacrificial material, such as silicon dioxide, are deposited and patterned. The sacrificial material acts as an intermediate spacer layer and is etched away to produce a freestanding structure. Surface micromachining technology allows smaller and more complex structures with multiple layers to be fabricated on a substrate. However, annealing or special deposition process control is required to reduce stresses in the layers that can cause warping. In contrast, bulk micromachining typically is stress free.

Surface micromachining has been used to manufacture an accelerometer for automotive air bag applications. A three-layer differential capacitor is created by alternate layers of polysilicon and phosphosilicate glass (PSG) on a

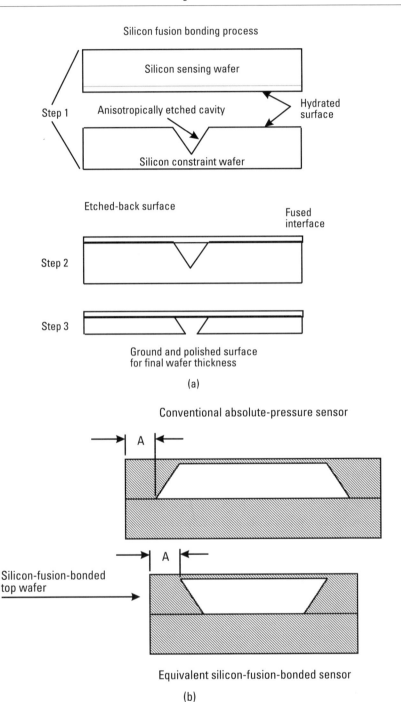

Figure 2.5 (a) Silicon fusion bonding process and (b) its effect on die size. (*After:* [12].)

Table 2.3
Comparison of Typical Wafer Bonding Process Conditions [8]

Process	Temperature (°C)	Pressure (Bar)	Voltage (V)	Surface Roughness (nm)	Precise Gaps	Hermetic Seal	Vacuum Level During Bonding (Torr)
Silicon-silicon (glass frit)	400–500	1	N/A	N/A	No	Yes	10
Anodic	300–500	N/A	100–1,000	20	Yes	Yes	10^{-5}
Silicon fusion	1,000	N/A	N/A	0.5	Yes	Yes	10^{-3}

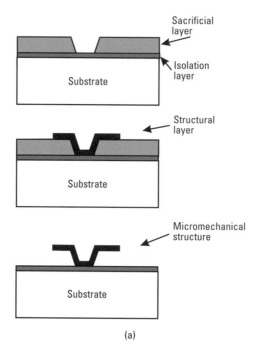

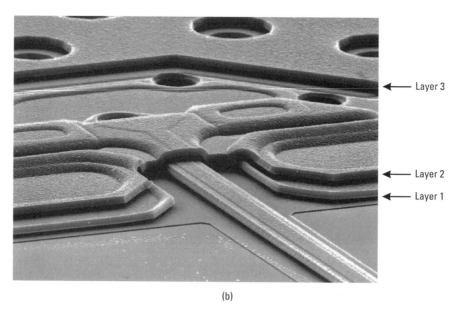

Figure 2.6 Surface micromachining: (a) process steps and (b) example of actual 3-layer structure.

4-inch (100-mm) wafer that is 0.015 inch (0.38 mm) thick [15]. A silicon wafer serves as the substrate for the mechanical structure. The trampoline-shaped middle layer is suspended by four support arms. That movable structure is the seismic mass for the accelerometer. The upper and lower polysilicon layers are fixed plates for the differential capacitor. The PSG is sacrificially etched by an isotropic etch, such as hydrofluoric acid (HF).

Because of the small spacing ($\approx 2\ \mu$m) possible with surface micromachining, new issues arise that affect both the sensor design and the manufacturing process. Squeeze-film damping, stiction, and particulate control must be addressed in each new design. The next three sections describe those areas and some approaches being used to deal with them.

2.4.1 Squeeze-Film Damping

The movement of structures separated by only a few microns can be greatly affected by the actual spacing and ambient (gas or vacuum) between the structures. That effect is known as squeeze-film damping. Squeeze-film damping can be significant in bulk micromachined capacitive structures, in which closer spacing is needed to achieve higher capacitance values. It is inherent in surface micromachining, in which spacing is only a few microns. For a particular structure, the gas that separates the layers has a viscous damping constant that increases with the inverse cube of the spacing [16]. Incorporating holes in the surface micromachined structure allows the damping to be tuned for desired characteristics. Holes also provide distributed access for the etchant to reduce etching time and the possibility of overetching portions of the structure.

2.4.2 Stiction

Stiction (static friction) is a phenomenon that occurs in surface micromachining resulting from capillary (van der Waals) forces generated during the wet etching of the sacrificial layers [17]. Under certain fabrication conditions, the microstructures can collapse and permanently adhere to the underlying substrate. The failure is catastrophic and must be prevented to achieve high-process yield and a reliable design. Preventing the top structure from contacting the bottom structure requires minimizing the forces acting on the device when the liquid is removed, or minimizing the attractive forces between the structures if they contact each other. Techniques used to prevent stiction depend on the manufacturer, the product design, and the process flow [17]. Many other solutions for stiction are being developed.

2.4.3 Particulate Control

One of the design problems that must be solved in working with structures separated by only a few microns is avoidance of contamination. Wafer-level packaging is an attractive solution because it provides a low-cost, protective, and safe environment for moving parts that require additional electrical testing and assembly processes. The package protects the device from microscopic particulates and handling and provides an ambient atmosphere for adjusting damping. A hermetically sealed accelerometer chip that can be overmolded in a low-cost conventional epoxy package is one example of an approach to avoiding contamination.

The general concept of a sensing die with moving parts that needs protection from the environment is shown in Figure 2.7 [15]. The three polysilicon layers are sealed inside a protective cavity formed by the silicon substrate, a bulk micromachined top (cover) wafer, and a glass layer that entirely surrounds the polysilicon structure. The glass is spaced a distance from the polysilicon structure to avoid the possibility of mechanical interference. The glass serves not only as the bonding medium, but also as the "mechanical" spacer that provides the elbow room for the movable structure. A 0.015-inch silicon wafer is used as the top or cover wafer. The glass is applied to the top wafer, which is then thermocompression-bonded to the bottom wafer, which contains the micromachined accelerometer structures. The top-wafer design provides a hermetic environment, physical protection, and access to the bond pads. When bonded, a sealed cavity for controlled squeeze-film damping is achieved.

Other techniques to minimize particle contamination use metal can or ceramic packaging. In those cases, the final package provides a hermetic environment for the structure. Prior to packaging, attention must be given to other processing steps that could allow particles to be trapped in the structure.

2.4.4 Combinations of Surface and Bulk Micromachining

Another combination of surface and bulk micromachining has been reported. An airgap capacitive pressure sensor has been demonstrated that combines bulk and surface micromachining on a single wafer [18]. The structure in Figure 2.8 used standard IC processing to create n-channel metal oxide semiconductor (NMOS) circuits with an additional polysilicon layer to produce a capacitor with a 0.6-μm-thick dielectric. Surface micromachining allowed a smaller gap to be produced. MOS circuitry on the top of the wafer was not exposed to the pressure medium. The inlet for the pressure source and the release for the surface micromachined structure were bulk micromachined into the silicon substrate using a KOH etch. A sensitivity of 0.93 mV/kPa (6.4 mV/psi) was

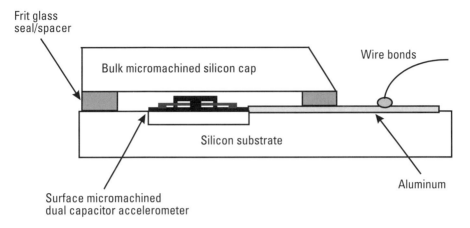

Figure 2.7 Combined surface and bulk micromachining in an accelerometer for protection from contamination.

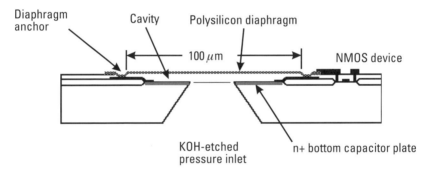

Figure 2.8 Combined surface and bulk micromachining in a pressure sensor. (*After:* [18].)

measured for the 100-μm by 100-μm capacitive structure. The 100-fF (femtofarad = 10^{-15}) capacitor had a resolution of 30 attofarads (10^{-18}).

2.5 Other Micromachining Techniques

The micromachining techniques discussed so far have been wet-etching techniques used primarily for piezoresistive and capacitive sensing for measuring pressure and acceleration. They are the basic processes used in research that have progressed into commercial manufacturing. However, other techniques are being developed that overcome the limitations, extend the types of measurements that can be made, or enhance the capabilities of previously discussed approaches. The newer techniques include the LIGA process, several methods of dry etching, micromilling, and laser micromachining.

2.5.1 LIGA Process

One of the newer micromachining processes is the LIGA (derived from German terms for *lithography, electroforming,* and *molding*) process, which combines X-ray lithography, electroforming, and micromolding techniques [19]. The LIGA process allows high-aspect-ratio (height/width) structures to be fabricated. X-ray–patterned photoresist molds are chemically etched in a metal plate. A polyimide layer a few microns thick acts as a sacrificial layer. A complementary structure is built up by electrodepositing a metal layer, such as nickel. After the final etching process, portions of the microstructure remain attached to the substrate and are able to move freely. Temperatures are under 200°C for the entire process. This process has made tiny 100-μm gears (about the diameter of a human hair). Up to a dozen metal gears have been driven by a low-level magnetic field powered micromotor. The LIGA process greatly expands micromachining capabilities, making possible vertical cantilevers, coils, microoptical devices, microconnectors, and actuators [14]. However, the LIGA process is difficult to integrate with electronics and has a high capital investment cost compared to bulk and surface micromachining. Table 2.4 compares other aspects of the three micromachining processes [20].

A differential pressure sensor with double-sided overload protection has been fabricated with a modified LIGA process [21]. As shown in Figure 2.9, this design combines isotropic bulk micromachining for the cavity and flow channels, LIGA processing for the electroplated nickel structure (which is 100-μm thick and has a gap of 0.80 μm), and sacrificial etching for the polysilicon diaphragm. The high-aspect-ratio metallic stop limits motion and suppresses diaphragm stress, as well as facilitates the option of a second signal to verify performance.

2.5.2 Dry-Etching Processes

Plasma etching and reactive ion etching can produce structures that are not possible from the wet chemical etching processes. Plasma etching is an etching process that uses an etching gas instead of a liquid to chemically etch a structure. Plasma-assisted dry etching is a critical technology for manufacturing ultra-large-scale integrated circuits [22]. Plasma-etching processes are divided into four classes: sputtering, chemical etching, energetic or damage-driven ion

Table 2.4
Comparison of Bulk, Surface, and LIGA Micromachining Processes (*After:* [20])

Capability	Bulk	Surface	LIGA
Planar geometry	Rectangular	Unrestricted	Unrestricted
Min. planar feature size	$2 \times$ depth	$1 \mu m$	$3 \mu m$
Sidewall features	54.74° slope	Dry etch limited	$0.2 \mu m$ runout over $400 \mu m$
Surface and edge definitions	Excellent	Mostly adequate	Very good
Material properties	Very well controlled	Mostly adequate	Well controlled
Integration with electronics	Demonstrated	Demonstrated	Difficult
Capital investment and cost	Low	Moderate	High
Published knowledge	Very high	High	Low

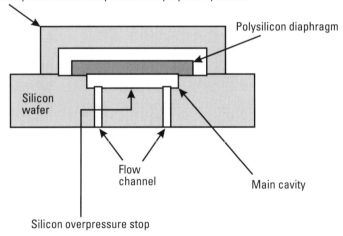

Figure 2.9 LIGA process combined with sacrificial etching for differential pressure sensor. (*After:* [21].)

etching, and sidewall inhibitor ion-assisted anisotropic etching. Figure 2.10 illustrates the different mechanical structures that result from these techniques.

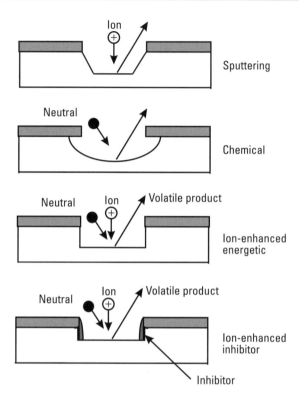

Figure 2.10 Different structures from plasma etching. (*After:* [22].)

Manufacturing control of a plasma-etching process addresses etch-rate control, selectivity control, critical-dimension control, profile control, and control of surface quality and uniformity. The control of these parameters allows new structural elements to be achieved. Plasma etching was used in the surface micromachined sensor in Figure 2.8. SF_6 was used to etch low-pressure chemical vapor deposition (LPCVD) silicon nitride and polysilicon; CF_4 was used to etch boro-phospho-silicate glass (BPSG); and CH_4 was used to etch BPSG and the nitride. Plasma etching was also performed on the aluminum metal [22].

Ion-beam milling is a dry-etching process that uses an ion beam to remove material through a sputtering action. It can be used separately or with plasma etching. When it is combined with plasma etching, it is also known as reactive ion etching (RIE).

The RIE process has been used for mechanical structures and capacitors in single-crystal silicon [23]. The RIE process was used to achieve high-aspect-ratio mechanical structures difficult to achieve using surface micromachining and wet-etch bulk micromachining. Structures with feature sizes down to

250 nm and with arbitrary structure orientations on a silicon wafer have been produced using the single-crystal reactive etching and metallization (SCREAM) process. As shown in Figure 2.11 [23], a thick-film deposition is not required. A silicon cantilever beam is formed with aluminum electrodes on both sides of the beam using silicon dioxide for insulation and for the top and

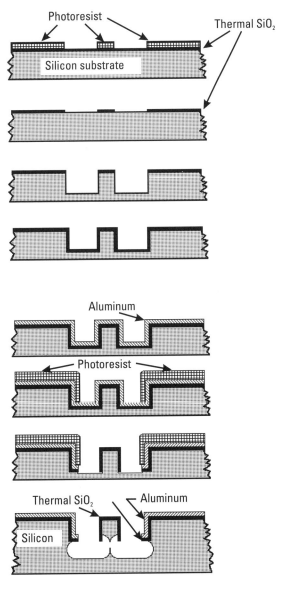

Figure 2.11 RIE in the SCREAM process. (*After:* [23].)

sidewall-etching mask. The SCREAM process can be used to fabricate complex circular, triangular, and rectangular structures in single-crystal silicon.

2.5.3 Micromilling

Milling at the microlevel can be performed using tools manufactured using another micromachining process. A focused ion beam (FIB) micromachining process has been used to precisely remove material from M42 cobalt high-speed steel with a Rockwell C hardness of 65–70 [24]. Four-fluted, two-fluted, and two-fluted square micromilling tools were manufactured with FIB micromachining. The micromilling tools were subsequently used to cut narrow (4μm), intricate trenches in polymethyl methacrylate (PMMA).

2.5.4 Lasers in Micromachining

In addition to chemical etching, lasers are used to perform critical trimming and thin-film cutting in semiconductor and sensor processing. The flexibility of laser programming systems allows their usage in marking, thin-film removal, milling, and hole drilling [25]. Lasers also provide noncontact residue-free machining in semiconductor products, including sensors. The precise value of the thin-film resistors in interface circuits is accomplished by interactive laser trimming. Interactive laser trimming at the die level for micromachined sensors has been used to manufacture high-volume, interchangeable, calibrated, and compensated pressure sensors since the mid-1980s [26].

Lasers have been used to drill through silicon wafers as thick as 0.070 inch (1.78 mm) with hole diameters as small as 0.002 inch (50.8 μm) [25]. For example, 0.005-inch (127 μm) holes spaced on 0.010-inch (254 μm) centers have been drilled into a wafer 0.015 inch (0.381 mm) thick. The hole diameters and close spacing are achieved without causing fracturing or material degradation. Also, lasers can vaporize the material (ablation) using high-power density.

Lasers have also been investigated as a means of extending the bulk micromachining process [27]. Figure 2.12 shows that either <110> or <100> wafers can be processed using a combination of photolithography, laser melting, and anisotropic etching. A deeper and wider etch occurs in the area that has been damaged. The grooved shape or microchannel obtained by this process has been used to precisely position fibers and spheric lenses in hybrid microoptical devices without requiring additional bonding or capturing techniques.

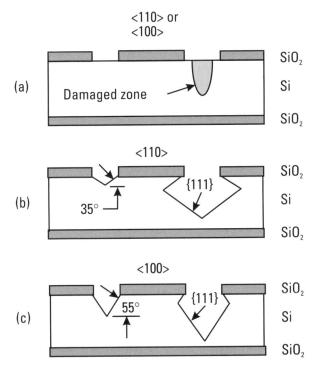

Figure 2.12 Combined effect of laser micromachining and anisotropic etching. (*After:* [27].)

2.5.5 Chemical Etching and IC Technology

While many of the processes used to manufacture a MEMS structure are similar to processes used to fabricate ICs, there are significant differences that make combining ICs and microstructures a nontrivial task. Table 2.5 shows the difference between IC and micromachining processes in five different areas [28]. One of the main concerns in integrating circuitry with micromachined structures is the effect that the etching and the resulting temperatures can have on other semiconductor processes and the circuit elements. The combined surface and bulk micromachining process shown Figure 2.8 also included NMOS circuitry to process the signal for the capacitive sensor. Other examples will be discussed to show the different approaches that are being pursued.

The process for an integrated sensor is more complex because it involves the joining of unique sensor processes with traditional IC processes. For example, bipolar operational amplifiers (op amps) typically are fabricated on <111> silicon substrates, while pressure sensors typically are fabricated on <100>

Table 2.5
Comparison of IC and Micromachining Processes (*After:* [28])

Device Type	Film Thickness (μm)	Critical Dimension (μm)	Aspect Ratio	Topography (μm)	Device Size (μm)
ICs	<1	0.35	2:1	<1	1
MEMS	2–6	1.00	6:1	4–10	100

silicon substrates. A production solution to that problem uses a standard op amp design on a <100> silicon substrate [29]. While that reduces the breakdown voltage in the op amp, it does not result in breakdown voltages less than the 10V required for the sensor. The d silicon substrate is p+-type with an n-type epitaxial layer. The n-type epitaxial layer matches both the op amp and the pressure sensor requirements. The pressure sensor also has several unique processes, such as cavity etch and a thin-film requirement, that must be combined with the op amp processing.

The task of combining two different process flows is complicated by allowances that must be made for each process. The final combination allows the diffusion of the sensor to remain the same and preserves the general process flow of the op amps. The cavity etch process defines and etches the silicon to produce a diaphragm in the silicon. The etch process calls for several masking steps to protect the silicon during the etch. The layers must be added into the flow in a manner that allows a proper layer structure.

The manufacturing steps for the integrated sensor include a thin-film process that serves as the circuit link between the pressure sensor and the op amp. The thin film is patterned to make laser-trimmable resistors. The relatively high resistance but low temperature coefficient of resistance (TCR) of the metal films have allowed resistor networks to be manufactured economically on a chip. Each unit is individually laser trimmed to the proper offset, sensitivity, and temperature characteristics. The use of those techniques enables robust, cost-effective, and precise integrated sensors to be manufactured in high volume. Figure 2.13 shows a scanning electron microscope (SEM) view of the single piezoresistive sensing element and circuitry that were integrated in the final process.

Standard commercial CMOS foundries, through the Metal Oxide Semiconductor Implementation System (MOSIS) service, provide the capability to have CMOS circuitry and post-processing EDP maskless etching to yield

Figure 2.13 Monolithic piezoresistive pressure sensor. (Courtesy of Motorola, Inc.)

integrated circuitry and mechanical structures on the same silicon wafer [30]. The micromachining is performed after the CMOS processing has been completed, which allows mixed electrical and mechanical silicon devices to be developed and even produced in small volumes. With that approach instrumentation and measurement devices can be reduced in both size and cost. High-volume, cost-sensitive applications, such as automobiles or appliances, will require a dedicated wafer-fabrication facility under the direct control of the sensor manufacturer.

Another standard 3-μm CMOS process has been combined with anisotropic wet etching at Fraunhofer Institute in Berlin [31]. The parameters of the CMOS process were not modified. Six more masking layers were needed to fabricate monolithic accelerometers. The sensor structure was defined in the CMOS processing, and testing of the IC portion was possible prior to the micromachining steps. Both top and backside (with etch stop) anisotropic etching were performed to create the accelerometer structure.

A modular CMOS and surface micromachining process was developed at the University of California [32]. A silicon foundry can perform the highly optimized CMOS process. The aluminum interconnects in the CMOS process were replaced with tungsten, and a titanium-nickel/titanium-silicon (TiN/ TiSi$_2$) diffusion layer at metal-silicon contacts was added to increase the temperature ceiling for the process. Oxide and nitride films were deposited to isolate the CMOS from the polysilicon.

Processing temperatures are a major concern for monolithic integrated sensors. With bulk micromachining, the same processing temperature constraints apply to both the sensor and the control electronics. With surface micromachining techniques, the structures are dielectrically isolated, so that leakage is not a problem at elevated temperatures. The maximum operating temperature is, therefore, limited by the control electronics.

Approaches for integration at the component level can be totally monolithic or multichip modules with separate die for sensor and the smart interface. The Microelectronics Center of North Carolina (MCNC) has investigated flip-chip technology (discussed in detail in Chapter 10) to add circuitry to processed micromachined devices with solder bumping [33]. That approach has been studied for surface micromachining and a combined surface and bulk micromachining process. Substrate materials, including silicon, Pyrex®, quartz, and gallium arsenide (GaAs), have been explored.

Die size and process complexity are basic drivers of cost for semiconductor processes and for sensors. The smaller the die size, the greater the number of die that fit on a particular wafer. The simpler the process, the lower the cost for processing and the higher the yield. Combined processes must achieve cost reduction, performance advantage(s), or improved reliability to be competitive in the marketplace.

2.6 Other Micromachined Materials

The most common IC materials [silicon (Si), silicon dioxide (SiO$_2$), aluminum (Al), and silicon nitride (Si$_3$N$_4$)] have played an important role in established micromachining technology. However, improving the performance of a sensor for a particular application, addressing a higher operating range, or sensing a new parameter can require other materials and sensing techniques. Table 2.1 compared GaAs, silicon carbide (SiC), and diamond to silicon. Silicon begins to exhibit plastic deformation at temperatures above 600°C. Higher operating temperatures are among the desirable properties that make those alternative materials attractive for ICs *and* semiconductor sensors. Other semiconductor materials, such as indium phosphide (InP), are being investigated for micromachined sensors. Also, several metals, metal oxides, and polymer films are being used in micromachined devices. The following four examples demonstrate the variety of approaches being pursued.

2.6.1 Diamond as an Alternative Sensor Material

Selectively deposited diamond film has been used as the thermal element in a flow sensor [34]. The bulk micromachined structure with diamond film is shown in Figure 2.14. Processing for the boron-doped heater resistors is performed at temperatures above 1,000°C. After surface preparation, the diamond film is grown in the desired regions by microwave plasma chemical vapor deposition in a mixture of hydrogen and methane gas and at a substrate temperature of 900°C. The aluminum-pad contact to the resistors is the next step. Last, the back of the wafer is anisotropically etched to form the bridge structure for the flow sensor. Besides higher operating temperatures, diamond's resistance to corrosive and abrasive environments makes it attractive as a flow sensor.

2.6.2 Metal Oxides and Piezoelectric Sensing

In addition to semiconductor materials, various metal oxides can be deposited on micromachined structures. A pressure sensor with a zinc oxide (ZnO) piezoelectric sensing element has been made in an IC-compatible process [35]. Surface micromachining is used to create the cavity. As shown in Figure 2.15, a thermal oxide layer (Tox) is grown to isolate the sensor from the silicon substrate. The lower polysilicon electrode is encapsulated in Si_3N_4. The spacer oxide layer is sacrificially etched. Polysilicon is used to form an electrically conductive structural support and cavity for the sensor. The 0.95-μm-thick active ZnO layer is deposited by RF sputtering. In addition to forming the active piezoelectric film, the sputtering also seals the sidewalls of the structure with an O_2-Ar mixture at 10 mtorr inside the cavity. The sensor exhibits approximately 0.36 mV/μbar sensitivity at 1.4 kHz and 3.4 dB variation over the range

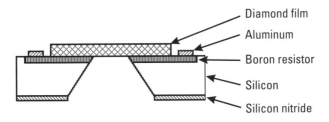

Figure 2.14 Diamond film flow sensor. (*After:* [34].)

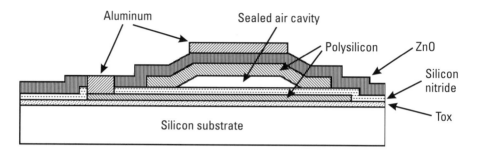

Figure 2.15 Piezoelectric surface micromachined sensor. (*After:* [35].)

200 Hz to 40 kHz. The dimensions of the sensor range from 50 by 50 μm^2 to 250 by 250 μm^2.

Surface micromachining techniques have been applied in a GaAs metal-semiconductor field effect transistor (MESFET) process to form both sensors and electronic circuits [36]. The sensors incorporate a 1-μm-thick sputtered ZnO capacitor supported by a 2-μm aluminum membrane formed on the GaAs substrate. The piezoelectric pressure sensor is 80 μm by 80 μm. Circuits using 4-μm GaAs MESFETs are fabricated in a planar, direct ion-implanted process. The sensors and circuits both operate at temperatures up to 200°C.

2.6.3   Films on Microstructures

The sensitive layers for chemical sensors can be deposited over a bulk micromachined structure. As shown in Figure 2.16, one approach to a semi-conductor gas sensor uses a thin metal oxide semiconductor deposited over an integrated heating element [37]. The resistivity of the metal oxide depends on the reducing or oxidizing environment being sensed. The sensitive area is thermally insulated from the silicon substrate to minimize power consumption. Response time of the device is less than 1 second with an operating temperature range of 250–400°C. The structure is contained in a chip 3.5 mm by 3.5 mm by 0.3 mm.

Polymer films are also being deposited on both surface and bulk micromachined structures. The films act as a membrane in biosensors. Multiple sites can have a variety of films sensitive to different levels of a particular substance or to different substances. These sensors call for a very small sample material to provide an analysis, making them ideal for usage on small children or any application where only a small sample is available.

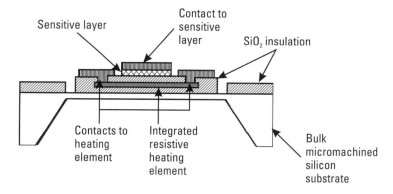

Figure 2.16 Sensing layer deposited over bulk micromachined structure. (*After:* [37].)

2.6.4 Micromachining Metal Structures

A micromachining process called electrochemical fabrication (EFAB) has been developed that can build a wide range of three-dimensional structures [38]. The process relies on the electroplating of metal and an in-situ mask patterning technique. The EFAB process is shown in Figure 2.17.

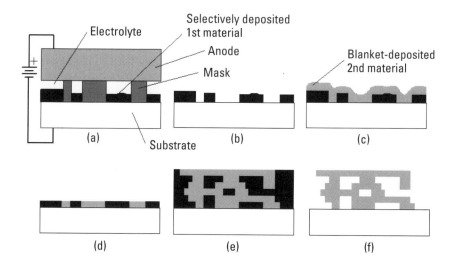

Figure 2.17 EFAB process. (Courtesy of USC/Information Sciences Institute.)

The first metal is deposited on a substrate using a conformable insulator patterned on an anode, as shown in Figure 2.17(a), to produce the partial layer of Figure 2.17(b). A second metal is blanket deposited over the first, Figure 2.17(c), and then the entire structure is mechanically planarized to a precise thickness and flatness, as shown in Figure 2.17(d). The process can be repeated for multiple layers. The final structure, such as Figure 2.17(e), is then etched to produce a three-dimensional device, Figure 2.17(f). The sacrificial metal is analogous to the sacrificial material in surface micromachining, except it is fully conductive. Therefore, an arbitrary patterning layer, including one with an overhanging structure, can be electrodeposited over it. The low process temperature makes it a candidate for additional CMOS circuit integration.

2.7 Summary

The trends in the sensing industry for micromachined structures include the following:

- Improved accuracy and resolution;
- Extended operating range of existing sensors;
- The development of higher operating temperature sensors;
- More complex structures using advanced micromachining technology and new materials;
- The addition of integrated circuits;
- Cost reduction through process improvements and die shrinking.

More complex structures will enable new parameters to be sensed and combined measurements to be performed. The continued merging of electronics and various micromachining techniques promises to be a key factor for smart sensing.

References

[1] Bryzek, J., K. Peterson, and W. McCulley, "Micromachines on the March," *IEEE Spectrum,* May 1994, pp. 20–31.

[2] http://mems.eeap.cwru.edu/SiC.

[3] Howe, R. T., et al., "Silicon Micromechanics: Sensors and Actuators on a Chip," *IEEE Spectrum,* July 1990, pp. 29–35.

[4] Goodenough, F., "Sensor ICs: Processing, Materials Open Factory Doors," *Electronic Design*, Apr. 18, 1985, pp. 130–148.

[5] Petersen, K., "Silicon as a Mechanical Material," *Proc. IEEE*, Vol. 70, No. 5, May 1982, pp. 420–457.

[6] Wise, K. D., "VLSI Circuit Challenges for Integrated Sensor Systems," *IEEE 1990 Symposium on VLSI Circuits*, pp. 19–22.

[7] Williams, K. R., and R. S. Muller, "Etch Rates for Micromachining Processes," *J. Micromechanical Systems*, Vol. 5, No. 4, Dec. 1996, pp. 256–269.

[8] Mirza, A. R., and A. A. Ayon, "Silicon Wafer Bonding: Key to MEMS High-Volume Manufacturing," *Sensors*, Dec. 1998, pp. 24–33.

[9] Frank, R., and J. Staller, "The Merging of Micromachining and Microelectronics," *Proc. 3rd Internat'l Forum on ASIC and Transducer Technology*, Banff, Alberta, Canada, May 20–23, 1990, pp. 53–60.

[10] Zhang, Y., S. B. Crary, and K. D. Wise, "Pressure Sensor and Simulation Using the CAEMEMS-D Module," *IEEE Solid-State Sensor and Actuator Workshop*, Hilton Head, SC, June 4–7, 1990, pp. 32–35.

[11] Behr, M. E., C. F. Bauer, and J. M. Giachino, "Miniature Silicon Capacitance Absolute Pressure Sensors," *Third Internat'l Conf. Automotive Electronics*, London, Oct. 2–23, 1981, pp. 255–260.

[12] Peterson, K., and P. Barth, "Silicon Fusion Bonding: Revolutionary Tool for Silicon Sensors and Microstructures," *Proc. Wescon '89*, pp. 220–224.

[13] "Strides Made in Wafer Bonding," *Electronic Engineering Times*, Jan. 4, 1999, p. 43.

[14] Ristic, L. J., *Sensor Technology and Devices*, Norwood, MA: Artech House, 1994.

[15] Ristic, L. J., et al., "A Capacitive Type Accelerometer With Self-Test Feature Based on a Double-Pinned Polysilicon Structure," *Digest of Technical Papers for Transducers '93*, June 1993, pp. 810–813.

[16] Suzuki, S., et al., "Semiconductor Capacitance-Type Accelerometer With PWM Electrostatic Servo Technique," *SAE Sensors and Actuators 1991 P-242*, Warrendale, PA, pp. 51–57.

[17] Mastrangelo, C. H., and C. H. Hsu, "Mechanical Stability and Adhesion of Microstructures Under Capillary Forces—Part I: Basic Theory and Part II: Experiments," *J. Microelectromechanical Systems*, Mar. 1993, pp. 33–55.

[18] Kung, J. T., and H. S. Lee, "An Integrated Air-Gap Capacitor Pressure Sensors and Digital Readout With Sub-100 Attofarad Resolution," *J. Microelectromechanical Systems*, Sept. 1992, pp. 121–129.

[19] *Microelectromechanical Systems Advanced Materials and Fabrication Materials*, Washington, D.C.: National Academy Press, 1997.

[20] Tang, W. C., "JPL Workshop Examines MEMS Reliability," *Micromachine Devices*, Vol. 2, No. 12, Dec. 1997, p. 4.

[21] Choi, B., et al., "Development of Pressure Transducers Utilizing Deep X-Ray Lithography," *IEEE 91CH2817-5 From Transducers '91*, pp. 393–396.

[22] Flamm, D. L., "Feed Gas Purity and Environmental Concerns in Plasma Etching," *Solid State Technology*, Oct. 1993, pp. 49–54.

[23] Zhang, Z. L., and N. C. McDonald, "A RIE Process for Submicron, Silicon Electromechanical Structures," *J. Micromechanics and Microengineering*, Mar. 1992, pp. 31–38.

[24] Friedrich, C. R., and M. J. Vasile, "Development of the Micromilling Process for High-Aspect-Ratio Microstructures," *J. Microelectromechanical Systems*, Vol. 5, No. 1, Mar. 1996, pp. 33–38.

[25] Swenson, E. J., "Laser Micromachining for Circuit Production," *Microelectronic Manufacturing and Testing*, Mar. 1990, pp. 17–18.

[26] Cumberledge, W., and J. M. Staller, "An Integrated On-Chip Pressure Sensor for Accurate Control Applications," *IEEE Solid-State Sensors and Actuators Workshop*, Hilton Head, SC, 1986.

[27] Alavi, M., et al., "Laser Machining for Fabrication of New Microstructures," *IEEE 91CH2817-5 From Transducers '91*, pp. 512–515.

[28] Romig, A. D., Jr., and J. H. Smith, "The Coming Revolution in ICs: Intelligent, Integrated Microsystems," *Micromachine Devices*, Vol. 3, No. 2, Feb. 1998, pp. 4–6.

[29] Baskett, I., R. Frank, and E. Ramsland, "The Design of a Monolithic Signal Conditioned Pressure Sensor," *IEEE 91CH2994-2 Custom Integrated Circuits Conference '91*, pp. 273.1–273.4.

[30] Marshall, J. C., et al., "High-Level CAD Melds Micromachined Devices With Foundries," *IEEE Circuits and Devices*, Nov. 1992, pp. 10–17.

[31] Riethmueller, W., et al., "Development of Commercial CMOS Process-Based Technologies for the Fabrication of Smart Accelerometers," *IEEE 91CH2817-5 From Transducers '91*, pp. 416–419.

[32] Yun, W., R. T. Howe, and P. R. Gray, "Surface Micromachined, Digitally Force-Balanced Accelerometer With Integrated CMOS Detection Circuitry," *IEEE Sensors and Actuators Workshop*, Hilton Head, SC, June 22–25, 1992, pp. 126–131.

[33] Markus, K. W., V. Dhuler, and A. Cowen, "Smart MEMS: Flip Chip Integration of MEMS and Electronics," *Proc. Sensors Expo*, Cleveland, Sept. 20–22, 1994, pp. 559–564.

[34] Ellis, C. D., et al., "Polycrystalline Diamond Film Flow Sensor," *IEEE Solid-State Sensors and Actuators Workshop*, Hilton Head, SC, June 4–7, 1990, pp. 132–134.

[35] Schiller, P., D. L. Polla, and M. Ghezzo, "Surface-Micromachined Piezoelectric Pressure Sensors," *IEEE 90CH2783-9 From Sensors and Actuators Workshop*, Hilton Head, SC, June 4–7, 1990, pp. 188–190.

[36] Choi, J. R., and P. Choi, "Micromachined ZnO Piezoelectric Pressure Sensors and Pyroelectric Infrared Detector in GaAs," *J. Electrical Engineering and Information Science*, Vol. 3, No. 2, Apr. 1998, pp. 239–244.

[37] Krebs, P., and A. Grisel, "A Low Power Integrated Catalytic Gas Sensor," *Sensors and Actuators B*, 13–14, 1993, pp. 155–158.

[38] Cohen, A. L., "3-D Micromachining by Electrochemical Fabrication," *Micromachine Devices*, Vol. 4, No. 3, Mar. 1999, pp. 6–9.

3

The Nature of Semiconductor Sensor Output

What a revoltin' development this is!
—Daffy Duck, from Warner Bros. cartoons

3.1 Introduction

Micromachining technology combined with semiconductor processing provides the ability to sense mechanical, optical, magnetic, chemical, biological, and other phenomena. These semiconductor sensors are, in most cases, a considerable improvement over the previous mechanical or other counterparts. However, as standalone components, they are far from the ideal characteristics that are desired for most measurements. This chapter analyzes the output from piezoresistive, capacitive, and piezoelectric sensors. Brief consideration is given for other sensing techniques and approaches to directly achieving a digital output from the sensor. The actual level of those signals and the various parameters used to specify the sensor's performance are discussed.

3.2 Sensor Output Characteristics

The sensing technique used for a particular measurement can vary considerably, depending on the range of the measurand, the accuracy required, environmental considerations that affect the packaging and reliability of the sensor, the dynamic nature of the signal, and the effect of other inputs on the measurand. The environmental considerations include operating temperature, chemical exposure, and media compatibility. The measurement systems also apply constraints. Factors such as signal conditioning, signal transmission, data display, operating life, servicing/calibration, impedance of sensor, impedance of system, supply voltage, frequency response, and filtering may make one solution preferable over another. In some cases, the sensor's design can make more than one sensing technology acceptable. That depends on the design and manufacturing capability of the sensor supplier. For example, ceramic and semiconductor sensors using capacitive or piezoresistive technology frequently compete for a particular application. Also, optoelectronic, Hall-effect, magnetoresistive, and even inductive sensors can measure displacement or velocity. However, once a technology has been accepted for a particular measurement, displacement by an alternative technology is difficult.

Determining the proper sensing technology for a particular application begins with understanding the fundamental design principle of the sensor and the specifications that the manufacturer guarantees. Semiconductor sensors have defined new terms and allow alternative design methodology for sensors based on the micrometer and nanometer scale in which they operate. This chapter uses piezoresistive pressure sensor examples to explain some key sensor parameters and a nontraditional design approach.

3.2.1 Wheatstone Bridge

The change in the resistance of a material when it is mechanically stressed is called *piezoresistivity*. Strain-gauge pressure sensors convert the change in resistance in four (sometimes only one or two) arms of a Wheatstone bridge, as shown in Figure 3.1(a). The output voltage of a four-element Wheatstone bridge, shown in Figure 3.1(b), is given by

$$E_o = \frac{E \Delta R}{R} \qquad (3.1)$$

where E_o is the output voltage, E is the applied voltage, R is the resistance of all bridge arms, and ΔR is the change in resistance due to an applied pressure [1]. Additional variable resistive elements are typically added to adjust the zero offset, calibrate sensitivity, and provide temperature compensation. The Wheatstone bridge can be operated in a constant voltage mode or a constant current mode. The constant voltage mode is more common because it is easier to generate a controlled voltage source. However, the constant current mode is also

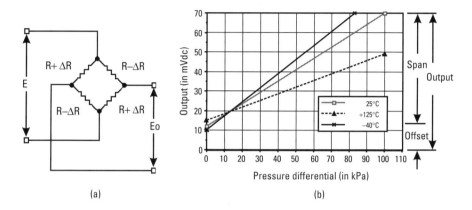

Figure 3.1 (a) Wheatstone bridge and (b) output curve for piezoresistive pressure sensor.

used and provides temperature compensation that is independent of the supply voltage.

Different approaches to piezoresistive strain gauges range from traditional bonded and unbonded to the newest integrated silicon pressure sensors. Pressure applied to the diaphragm produces a change in the dimension of the diaphragm, an increase in the length of the gauge (l), and a change in its resistance ($R = rl/A$, where R is the resistance of the gauge, r is the resistivity of the material, l is the length of the resistive element, and A is the cross-sectional area). The change in length per unit length is the strain. The sensitivity of a strain gauge is indicated by the gauge factor (GF), which is defined by:

$$K = \frac{\Delta R / R}{\Delta l / l} \tag{3.2}$$

where $K = \pi \cdot \varepsilon$ = the gauge factor, π is the piezocoefficient, and ε is the modulus of elasticity for the material.

K for metals is 2; for silicon, it is approximately 100 to 200, depending on the doping level and the design [2]. In contrast to metals, the change in resistance is a secondary effect for silicon. The primary effect is the change in conductivity of the semiconductor material due to its dependence on mechanical stress.

The GF for a strain gauge is improved considerably (to about 150) by using a silicon strain gauge. However, besides the conventional resistive Wheatstone bridge, silicon processing techniques and the miniature size of piezoresistive elements in silicon enable the design of a unique piezoresistive sensor. This alternative design demonstrates one of the different possibilities that semiconductor technology brings to sensing.

3.2.2 Piezoresistivity in Silicon

The analytic description of the piezoresistive effect in cubic silicon can be reduced to two equations that demonstrate the first-order effects [3].

$$\Delta E_1 = P_0 \cdot I_1(\pi_{11}X_1 + \pi_{12}X_2) \tag{3.3}$$

$$\Delta E_2 = P_0 \cdot I_2 \cdot \pi_{44}X_6 \tag{3.4}$$

where ΔE_1 and ΔE_2 are electric field flux density, P_0 is the unstressed bulk resistivity of silicon, I is the excitation current density, π is the piezoresistive coefficient, X_1 and X_2 are axial stress tensors, and X_6 is a shear stress tensor due to the applied force.

The effect described by (3.3) is utilized in a silicon pressure transducer of the Wheatstone-bridge type. Regardless of whether the sensor designer chooses n-type or p-type layers for the diffused sensing element, the piezoresistive coefficients π_{11} and π_{12} in (3.3) will have opposite signs. That implies that, through careful placement, orientation with respect to the proper crystallographic axis, and a sufficiently large aspect ratio for the resistors themselves, it is possible to fabricate resistors on the same diaphragm that both increase and decrease, respectively, from their nominal values with the application of stress.

The effect described by (3.4) is typically neglected as parasitic in the design of a Wheatstone-bridge device. A closer look at (3.4) reveals that the incremental electrical field flux density, ΔE_2, due to the applied stress, X_6, is monotonically increasing for increasing X_6. In fact, (3.4) predicts an extremely linear output, since it depends on only one piezoresistive coefficient and one applied stress. Furthermore, the incremental electric field can be measured by a single stress-sensitive element. That is the theoretical basis for the design of the transverse voltage or shear stress piezoresistive strain gauge.

Figure 3.2 shows the construction of a device that optimizes the piezoresistive effect of (3.4). The diaphragm is anisotropically etched from a silicon substrate. The piezoresistive element is a single, four-terminal strain gauge located at the midpoint of the edge of the square diaphragm at an angle of 45 degrees. The orientation of 45 degrees and the location at the center of the edge of the diaphragm maximizes the sensitivity to shear stress (X_6) and the shear stress being sensed by the transducer by maximizing the piezoresistive coefficient, π_{44}.

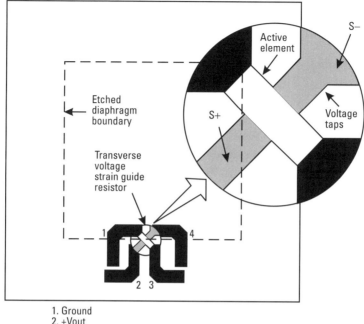

1. Ground
2. +Vout
3. Vs
4. −Vout

Figure 3.2 Shear stress strain gauge.

Excitation current is passed longitudinally through the resistor (pins 1 and 3), and the pressure that stresses the diaphragm is applied at a right angle to the current flow. The stress establishes a transverse electric field in the resistor that is sensed as an output voltage at pins 2 and 4, which are taps located at the midpoint of the resistor. The single-element shear stress strain gauge can be viewed as the mechanical analog of a Hall-effect device.

Using a single element eliminates the need to closely match the four stress- and temperature-sensitive resistors on the Wheatstone-bridge designs while greatly simplifying the additional circuitry necessary to accomplish calibration and temperature compensation. The offset does not depend on matched resistors but on how well the transverse voltage taps are aligned. That alignment is accomplished in a single photolithography step, making it easy to control, and is only positive, simplifying schemes to zero the offset. The temperature coefficient of the offset is small (nominally ±15 μV/°C), because multiple resistors and temperature coefficients do not have to be matched. By using proper doping levels, the temperature dependence of full-scale span (the difference between full-scale output and offset) can be carefully controlled and therefore compensated without the requirement of characterizing each device over the temperature range.

A more efficient single-element transducer design was developed recently to produce a higher output signal from the same level of mechanical stress [4]. Figure 3.3(a) shows the original X-ducer™ design and its location relative to the diaphragm edge. The design provided a tighter offset distribution, which was a significant improvement over Wheatstone-bridge designs used in the early days of micromachined pressure sensors. However, improvements in processing and design can achieve a tighter distribution without using the X-ducer™ design. As a result, the "picture frame" design shown in Figure 3.3(b) was developed. The picture frame design provides about 40% more span than the X-ducer™. That more efficient design allows both better linearity and higher signal-to-noise ratio. That improvement in output signal allows a lower overall gain to be used, enhancing accuracy.

3.2.3 Semiconductor Sensor Definitions

The batch-processing techniques used by semiconductor manufacturers are ideally suited for making high-volume, low-cost sensors. One limitation of batch-processed parts, however, is that certain parameters are not precisely specified but only listed as "typical" on the manufacturer's data sheet. In many cases, meeting fixed limits on all parameters can drive costs up and offset the benefits of high-volume batch processing [5].

Use of typical specifications is not necessarily a drawback, however. Uniformity in each wafer lot is a specific strength of semiconductor processing.

Piezoresistive sensing element

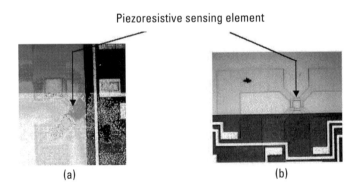

(a)   (b)

Figure 3.3 Piezoresistor design comparison of (a) original single element to (b) new design. (Courtesy of Motorola, Inc.)

Nevertheless, that factor must be considered to avoid problems in volume production.

The term *typical specification* generally has several implications in the semiconductor industry. Usually, it indicates a parameter that has been characterized during the design phase and represents the mean value for the manufacturing process. Certain manufacturers also add that a ±3 sigma value is the total spread that users can expect on typical specified parameters. That spread cannot be assumed, however, because *typical* means that limits are not normally attached to the manufacturer's quality assurance program.

In many cases, *typical* is used to indicate that the measurement process contributes more to the inaccuracy of the reading than to the actual variation from unit to unit. That practice shows why users must evaluate the parameter specified as typical. Pressure hysteresis in silicon diaphragms, for example, is essentially nonexistent and should not cause great concern if specified as typical. On the other hand, the variation over temperature of span and offset compensation could be significant and requires further investigation if listed as a typical specification.

Many parameters on sensor data sheets are specified as the plus-or-minus percentage of full-scale (output or span). That designation is the key to understanding other specifications. Full-scale output is the output at rated pressure including the zero offset. Full-scale span is the difference between the output with no pressure applied (offset) and the output at rated pressure. For a given device, for example, Motorola's MPX100, a 0.25% full-scale span rating on linearity can range from 0.113 mV for a low-output device (45 mV) to 0.225 mV for a device with the highest output (90 mV).

Sensitivity is the span divided by the operating pressure range. The sensitivities for devices made by various manufacturers are frequently compared. The pressure ranges and the supply voltage must be considered for an accurate comparison. The procedure often reveals interesting information that may affect a design decision, especially if the linearity at a given sensitivity is also taken into account.

The linearity or deviation from a straight-line relationship is expressed as a percentage of full-scale output (or span) and is also subject to different test methods. The least-squares, best-fit technique is the commonly accepted choice for establishing linearity. However, it requires several measurement points to provide acceptable results.

A simpler measurement is the end-point method, which requires only three points: zero, midscale, and full scale. The method lends itself to high-volume testing. Because it always yields a value that is about double the magnitude of a least-squares method, it is frequently more appropriate for error budget calculations. Any comparison of competitive data sheet values for linearity should take into account the method used to determine this parameter.

The effect of temperature on offset and span (and sensitivity) is one of the most critical and frequently most troublesome aspects of using semiconductor pressure sensors. Products that have no temperature compensation can be used with little effort for a narrow temperature range, such as 25°C ± 15°C or less, and accuracy requirements of a few percentage points. Certain sensors, such as Motorola's MPX family, can be temperature compensated to 1–3% accuracy over a 0–85°C temperature range without exercising the sensor over temperature. Ease of compensating for temperature is an important factor in evaluating a pressure sensor.

To accurately compare various products, similar units must be used. Full-scale value (F.S.), temperature range, and actual limits (specified as plus-or-minus percentage points F.S., plus-or-minus millivolts, or plus-or-minus percentage points F.S. per degree Celsius) can vary from one manufacturer to another. The conversion of those units to millivolts per degree Celsius can provide a good comparison. For example, a calibrated pressure sensor, such as Motorola's MPX2000 family with 40-mV output and ±1% F.S. variation over 0–85°C, has a maximum window of 9.4 μV/°C. That type of specification can indicate improved performance over reduced temperature spans. However, a user must know actual curve shapes within the temperature compensation window to have confidence in that value.

Three other terms on sensor data sheets are often a source of confusion: *static accuracy*, *ratiometricity*, and *overpressure*. Frequently, manufacturers lump together linearity, pressure hysteresis, and repeatability as static accuracy. Sometimes manufacturers indicate that characteristic, stated simply as the accuracy, as a single parameter on the data sheet.

Ratiometricity is one of the new terms that semiconductor manufacturers have brought to pressure sensing. Although output is given at a specific voltage rating, lower or higher voltage supplies can be used (within the maximum rating of the device) with constant voltage source sensors. Ratiometricity error is the change in the output resulting from a change to the supply voltage and is usually expressed as a percentage of full-scale output. The output in that instance varies as the ratio of source voltage to the manufacturer's rated voltage. A constant voltage is required for proper operation, a fact that should not be overlooked when considering the total circuit requirements.

The term *overpressure* takes on a different meaning with semiconductor pressure sensors. Even though the silicon diaphragm is only 0.001 inch (25.4 μm) thick, a device rated at 30 psi can withstand 200–300 psi without damage. Most manufacturers specify overpressure conservatively at two or three times the rated pressure.

Obviously, no damage should occur when the sensor operates within the overpressure rating. However, readings made above the normal range are another matter. Normally, linearity starts to degrade and could fall outside the specified rating if the sensor is operated above its rated pressure. If linearity were only marginally acceptable to begin with, overscale readings could cause trouble. However, that is not the case for many applications.

The normal mindset for making pressure readings with a pressure gauge is to avoid operating at full scale for best accuracy. Full-scale operation can also pin the gauge needle, resulting in miscalibration or damage. However, operating a semiconductor sensor at full scale usually provides better accuracy than a midscale reading. That is especially true for units measured for endpoint linearity, because output is accurately measured to specification at full scale. Slight excursions over the pressure rating do not significantly degrade linearity. However, if the output saturates above the rated pressure or the digital gauge's operating range is exceeded, a higher pressure input will not produce a higher output reading.

3.2.4 Static Versus Dynamic Operation

The protective gels or coatings that protect the active surface from the pressure media degrade response time for semiconductor pressure sensors. Also, isolating the semiconductor by stainless steel diaphragms and oil-filled chambers can decrease response time. However, response time within 1 ms is typically achieved when the sensor is exposed to a full-scale pressure excursion.

Measuring systems that operate at 6,000 rpm require sensors to operate at frequencies above 100 Hz. Frequently, a higher frequency range or higher frequency signal components are also of interest, as in acceleration and vibration sensors. However, lower frequency sensors tend to have lower noise floors. The lower noise floor increases the sensor's dynamic range and may be more important to the application than higher frequency capability [6].

3.3 Other Sensing Technologies

Piezoresistive sensing is the most common micromachined sensor because its output signal is predictable and an easy-to-signal condition. Other techniques are gaining in popularity, especially as sensing techniques for smart sensors are based on the capability of electronics circuitry to handle the signal conditioning. A brief review focusing on the output of current semiconductor sensing techniques follows.

3.3.1 Capacitive Sensing

Capacitive sensors typically have one plate that is fixed and one that moves as a result of the applied measurand. The nominal capacitance is $C = A\varepsilon/d$, where A is the area of the plate, ε is the dielectric constant, and d is the distance between the plates. Two common capacitive pressure sensors used in automotive applications are based on silicon and ceramic capacitors. A differential capacitor structure can also be used for acceleration. Figure 3.4 shows the surface micromachined capacitor and the resulting output [7].

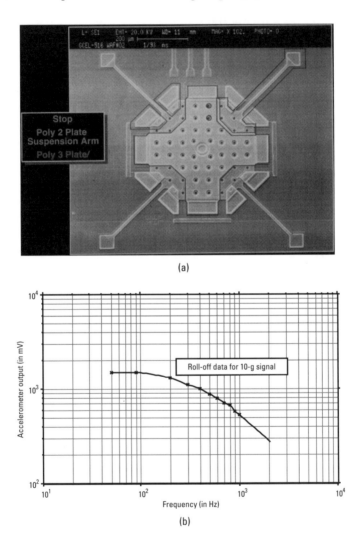

Figure 3.4 (a) Differential capacitor structure in accelerometer and (b) output.

Silicon technology combined with surface micromachining has allowed the capacitance between interdigitated fingers to measure acceleration and other inputs. The value of the nominal capacitance is in the range of 100 fF to 1 pF and the variation in capacitance is in femtofarads. Circuitry integrated on the same silicon chip or in the same package (that takes into account the parasitic capacitance effect of packaging) can use that level of signal to provide a useful output for a control system [8].

3.3.2 Piezoelectric Sensing

A piezoelectric sensor produces a change in electrical charge when a force is applied across the face of a crystal, ceramic, or piezoelectric film. The inherent ability to sense vibration and the necessity for high-impedance circuitry are taken into account in the design of modern piezocrystal sensors. Transducers are constructed with rigid multiple plates and a cultured-quartz sensing element that contains an integral accelerometer to minimize vibration sensitivity and suppress resonances.

Lead zirconate titanate (PZT) ceramic is used to construct biomorph transducers that sense motion, vibration, or acceleration and can be activated by an applied voltage. The biomorph consists of two layers of different PZT formulations that are bonded together in rectangular strips or washer shapes. The rectangular shape can be mounted as a cantilever. Motion perpendicular to the surface generates a voltage [9].

More recently, piezofilm sensors that produce an output voltage when they are deflected provide a method for inexpensive measurements. Figure 3.5(a) shows the construction of a piezoelectric film sensing element [10]. The polymer film generates a charge when it is deformed and also exhibits mechanical motion when a charge is applied to it. For a vehicle detector, the sensor is mounted in an aluminum channel filled with polyurethane and embedded in the pavement. The output of the sensor for three different types of vehicles is shown in Figure 3.5(b). In the voltage mode, both the weight and the speed of the vehicle passing over it influence the voltage. A positive output occurs when the tire compresses the film. The negative output results from expansion after the tire has passed over the sensor.

Surface micromachining techniques have been combined with piezoelectric thin-film materials, such as zinc oxide, to produce a semiconductor piezoelectric pressure sensor. One design has a sensitivity of 0.36 mV/μbar at 1.4 kHz [11]. Low-level acoustic measurements are a potential application for the technology.

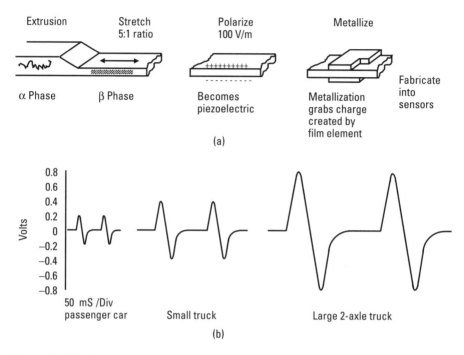

Figure 3.5 (a) A Kynar piezoelectric film is polarized in an intense electric field of approximately 100 V/μm and metallized to create a transducer. (b) The output of a vehicle sensor in a voltage mode is a voltage proportional to weight and speed of the vehicle. (*After:* [10].)

3.3.3 Hall Effect

A vertical Hall-effect structure has been designed for the detection of magnetic fields oriented parallel to the plane of the chip. Bulk micromachining was used to achieve higher sensitivity than a device without micromachining. The micromachined unit had an output of 70 mV for an applied magnetic field of 400 mT, which was almost five times the sensitivity of the unetched unit [12].

3.3.4 Chemical Sensors

A metal oxide chemical sensor's resistivity changes, depending on the reducing or oxidizing nature of the gaseous environment around the sensor. Figure 3.6 shows sensitivity data for hydrogen (H_2) and carbon monoxide (CO) in air for one design [13]. Conductances of over $100(10^{-7})\Omega^{-1}$ for H_2 and almost $32(10^{-7})\Omega^{-1}$ for CO were measured with the sensor exposed to concentration of

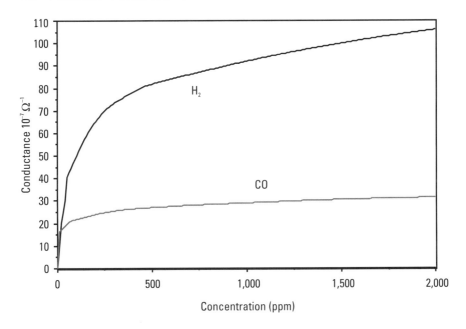

Figure 3.6 Conductance change of chemical sensor based on specific gas and concentration.

2,000 ppm of each gas in separate measurements and operating at a minimum temperature of 250°C.

The Jet Propulsion Laboratory (JPL) is investigating a hydrocarbon sensor in which a porous layer of SiC on top of 6H silicon carbide forms the channel of a field-effect transistor (FET) [14]. Ions produced during catalysis of a given hydrocarbon collect in the porous layer under the gate, affecting the current-voltage relationship of the FET. Circuitry scans the FET at a range of voltages and identifies a particular hydrocarbon compound. The intent is to integrate the sensing functions with additional electronics for an integrated probe for detecting hydrocarbons. The SiC sensor should be able to operate at temperatures as high as 500°C.

3.3.5 Improving Sensor Characteristics

The low-level output of the sensors described in this section and the design parameters discussed in Section 3.2.3 are among the characteristics that must be considered in the use of a semiconductor sensor in a particular application. Table 3.1 summarizes sensor characteristics and the design areas that are commonly used to improve those characteristics [15]. Improved sensor

Table 3.1
Common Undesirable Characteristics of Semiconductor Sensors (*After:* [15])

Characteristic	Sensor Design	Sensor Interface	MCU/DSP
Nonlinearity	Consistent		Reduce
Drift	Minimize		Compensate
Offset		Calibrate	Calibrate/reduce
Time dependence of offset	Minimize		Auto-zero
Time dependence of sensitivity			Auto-range
Nonrepeatability	Reduce		
Cross-sensitivity to temperature and strain		Calibrate	Store value and correct
Hysteresis	Predictable		
Low resolution	Increase	Amplify	
Low sensitivity	Increase	Amplify	
Unsuitable output impedance		Buffer	
Self-heating	Increase Z		PWM technique
Unsuitable frequency response	Modify	Filter	
Temperature dependence of offset			Store value and correct
Temperature dependence of sensitivity			Store value and correct

performance can result when the design of the sensor and the capability of subsequent components that achieve a smart sensor are taken into account in the overall design of the smart sensor. Digital logic provided by either an MCU or a DSP plays a vital role in smart sensing and in improving the sensor's performance.

3.4 Digital Output Sensors

A sensor that directly interfaces to an MCU without requiring A/D conversion simplifies system design and reduces the cost of the MCU. Today's sensors accomplish that with additional circuitry. However, the industry's quest is a sensor or sensor family with an inherent digital output. On-chip or external signal conditioning may achieve "inherent" capability, but it must cost considerably less and be more accurate than units using existing techniques.

3.4.1 Incremental Optical Encoders

Structures that contain a light source and a photodetector provide a digital means of measuring displacement or velocity when an alternating opaque and translucent grid is passed between them. As shown in Figure 3.7, a macrosized example of uniformly spaced apertures on a wheel allows logic circuitry to count the number of pulses in a given time frame to determine shaft velocity or angular displacement [16]. Linear measurements of displacement and velocity also are possible.

Basic presence sensing can be accomplished through the use of a single optical channel (or emitter-detector pair). Speed and incremental position sensing must use two channels. The most commonly used approach is quadrature sensing, in which the relative positions of the output signals from two optical channels are compared. That comparison provides the direction information, and either of the individual channels gives the transition signal to derive either count or speed information. A typical application can use the direction output as the up/down input and either channel A or channel B as the count input for a common up/down counter. The counter in an incremental system will increment or decrement as required to maintain a relative position or count output.

Quadrature direction sensing requires that the two optical channels create electrical transition signals that are out of phase with each other by 90 degrees nominally. When the code wheel is spinning, the two electrical output signals, A and B, will be 90 degrees out of phase, because the wheel windows are 90 degrees out of phase with the two sensor channels. The interpretation of

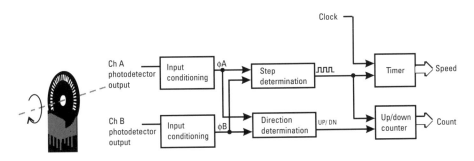

Figure 3.7 Incremental optical encoder.

whether signal A leading B is clockwise or counterclockwise is a matter of choice. Figure 3.8 shows the ideal quadrature outputs. The two waveforms are in four equal quadrants, exactly 90 degrees out of phase with each other. The detector is on when light is present and turns off when the web blocks the beam.

3.4.2 Digital Techniques

A study of available digital techniques has been reported [15]. Sensing techniques based on a resonant structure or a periodical geometric structure are potentially the most direct approach to digital sensing. However, other techniques can be used as well.

Micromachined silicon can act as a resonant structure if it is designed with a membrane or tuning fork. Electric activation is achieved by using a piezoelectric film such as ZnO on the micromachined structure. Most resonant structures are implemented in pairs. One element is not interfaced to the input and acts as a reference. Comparison of the output frequency of the sensing element to the reference element reduces the influence of unwanted parameters. The technique is commonly used in chemical sensors using surface acoustical wave (SAW) delay line oscillators.

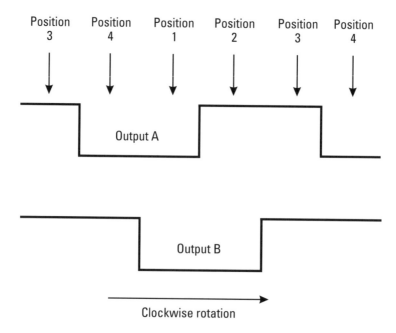

Figure 3.8 Quadrature detection.

Other approaches for digital output sensors include electrical oscillator-based (EOB) sensors and stochastic analog-to-digital (SAD) converters [15]. An EOB sensor is designed to generate a periodic current or voltage signal when it is subjected to a measurand. Current-to-frequency or voltage-to-frequency conversion provides a digital signal. The current-to-frequency technique has been demonstrated on a piezoresistive silicon pressure sensor. The frequency changed from 200–230 kHz with a pressure change from 0–750 mmHg.

A different approach for an EOB sensor was used with a capacitive sensor. Two equal but opposite current sources applied a square wave to a capacitive silicon pressure sensor. The zero pressure frequency was 155 kHz, and the sensitivity was 25 kHz/450 mmHg for that design.

A ring oscillator also has been used for EOB sensors. An odd number of integrated-injected logic (I^2L) gates connected in a ring on a piezoresistive pressure sensor provided the ring oscillator. Sensitivities up to 10.6 kHz/bar were obtained with a ring oscillator frequency of 667 kHz.

SAD converters employ the noise in flip-flop circuits to generate a random signal and use the flip-flop as a comparator. A piezoresistive sensor element in the flip-flop circuit is one way to implement this approach for a sensor. The number of 1s and 0s is a direct measure of the strain in a pressure sensor and a simple counting procedure for an MCU. A flip-flop sensor that can measure stresses as small as 8 kPa in a silicon cantilever beam has been reported.

A final digital example is provided by arrays of sensing elements [17]. A pressure switch with four switch points was fabricated using silicon fusion bonding. Pressure-switch points of $\frac{1}{4}$ (P1), $\frac{1}{2}$ (P2), $\frac{3}{4}$ (P3), and 1 (P4) atmosphere were designed using controlled thinning of the diaphragm. The switches close electrical contacts when their desired pressure threshold is exceeded. Table 3.2 shows the truth table for the design. The outputs can be directly applied to a logic control.

3.5 Noise/Interference Aspects

The general model for a transducer shown in Figure 1.2 indicated the interference sources that affect the low-level output of the sensor. An actual sensor signal combined with noise and interference from other sources, such as temperature, humidity, or vibration, can be a major problem when dynamic signals are measured. Compared to static signals where filtering can be used to minimize noise, dynamic measurements can pose a challenge [18].

Piezoresistive pressure sensors have two dominant types of noise [19]. Shot noise is the result of the nonuniform flow of carriers across a junction and

Table 3.2
Truth Table for Array of Four Pressure Switches

Pressure (P)	S1	S2	S3	S4	Out1	Out2	Out3	Out4
$P < P1$	Open	Open	Open	Open	0	0	0	0
$P1 \leq P < P2$	Closed	Open	Open	Open	1	0	0	0
$P2 \leq P < P3$	Closed	Closed	Open	Open	1	1	0	0
$P3 \leq P < P4$	Closed	Closed	Closed	Open	1	1	1	0
$P4 \leq P$	Closed	Closed	Closed	Closed	1	1	1	1

is independent of temperature. Shot noise increases in sensors with increased levels of integration. The second type of noise, flicker noise, or 1/f, results from crystal defects and wafer processing. Flicker noise is proportional to the inverse of frequency and is more dominant at lower frequencies. Decoupling and filtering can reduce these types of noise. Thermal or Johnson noise produced by thermal agitation of electronics in conductors and semiconductors is usually small in those devices.

3.6 Low-Power, Low-Voltage Sensors

Portable sensing applications, including portable data loggers and data acquisition systems, require sensors to operate at low power levels and low battery supply voltages. For low-power pressure measurements, high-impedance pressure sensors and pulse-width modulation (PWM) techniques have been developed to reduce power drain. Depending on the type of measurement (static or dynamic) and how frequently it must be measured, the sensing system can be in a sleep mode and wake up to make periodic readings that can be transmitted to distant recording instruments. This technique is useful in process controls, hazardous material monitoring systems, and a variety of data acquisition applications that would have been more time consuming, dangerous, prohibitively expensive, or too heavy with previous technology. Monitoring pressure is one of the more frequent measurements that must be made in systems, ranging from tire pressure on vehicles to leakage monitoring in tanks, blood pressure in portable healthcare monitors, and barometric pressure in weather sondes. The combination of higher impedance sensors with new design approaches, such as PWM input and other power management techniques, is increasing the applications in this area of portable equipment.

3.6.1 Impedance

Many mechanical transducers have a low impedance. Strain gauges, for example, are typically 350Ω. That impedance has been achieved in semiconductor devices, especially where direct replacement of a mechanical unit is required. In semiconductor circuits, low impedance is desired for noise purposes. However, high impedance is required to minimize the current draw for portable applications, to use existing interface circuits, and to prevent loading on amplification stages [20]. For high-impedance pressure sensors, input impedance in the range of approximately 5 kΩ is common. Output impedance for those devices is also approximately 5 kΩ. The sensor must be designed to achieve the results that the particular system requires.

3.7 Analysis of Sensitivity Improvement

Low-pressure measurements (< 0.25 kPa or 0.036 psi) are good examples of the problems that can occur and the solutions that have been developed to cope with the low-level signal. Low-pressure measurements are limited by existing silicon sensor designs. The sensitivity (and maximum stress) of a silicon diaphragm is directly related to the area and inversely related to the square of the thickness of the diaphragm. Three different approaches have been pursued to improve sensitivity.

3.7.1 Thin Diaphragm

One approach for increasing the sensitivity for low-pressure measurements is thinning the diaphragm. The typical thickness for silicon diaphragms ranges from 2 to 12 μm. However, a thin diaphragm can have unacceptable linearity. A number of researchers have investigated a variety of stress concentrators or bosses designed into the diaphragm structure to minimize the nonlinearity. The bosses provide a locally stiffer structure and limit the overall deflection of the diaphragm. A thinner diaphragm (4 μm), shallow resistor geometries in the submicron area, and advanced silicon micromachining have been used in a piezoresistive sensor to achieve a sensitivity of 50 mV/V/psi (7.3 mV/V/kPa) [21].

3.7.2 Increased Diaphragm Area

A second approach to low sensitivity involves increasing the sensitivity by increasing the size of the diaphragm. An area increase of over 2.5 times results in a 0.5-psi sensor with 7 mV/V/psi [22]. The larger area does not compromise ruggedness, but it does increase the cost of the sensor.

3.7.3 Combined Solution: Micromachining and Microelectronics

A more recent solution to increasing the sensitivity uses the combined capability of a low-pressure sensor and a microcontroller. Figure 3.9 shows the block diagram [23]. The microcontroller provides a signal to pulse the sensor with a higher supply voltage. The microcontroller also provides signal averaging for noise reduction and samples the supply voltage to reject sensor output variations due to power supply variations. An important factor for sensors making very low pressure measurements is stress isolation for the package. A piston-fit package for the unit used in this circuit allows 40 μV/Pa with a 15V source.

3.8 Summary

This chapter showed the distinct advantages that semiconductor sensors have over mechanical predecessors. However, these sensors require additional effort to obtain their full benefits in systems. Semiconductor sensors draw from

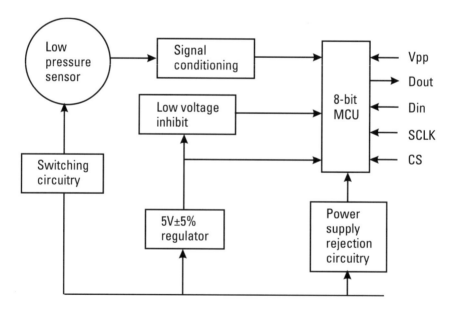

Figure 3.9 Low-pressure sensor combined with MCU to achieve higher pressure sensitivity.

previous mechanical sensors, but the physics of the microscale can provide alternative solutions that are not possible at the macrolevel.

References

[1] Carr, J. J., *Sensors and Circuits*, Englewood Cliffs, NJ: Prentice-Hall, 1993.

[2] Shankland, E. P., "Piezoresistive Silicon Pressure Sensors," *Sensors*, Aug. 1991, pp. 22–26.

[3] Frank, R. K., and W. E. McCulley, "An Update on the Integration of Silicon Pressure Sensors," *Wescon/85 Integrated Sensor Technology Session Record 27*, San Francisco, 1985, pp. 27/4–1–5.

[4] Bitko, G., R. Frank, and A. McNeil, "Improvements From Next-Generation MEMS Sensor Design" (to be published).

[5] Frank, R., "Designing With Semiconductor Pressure Sensors," *Machine Design*, Nov. 21, 1985, pp. 103–107.

[6] Barrett, R., and F. Wicoxon, "Monitoring Vibration With Piezoelectric Sensors," *Sensors*, Aug. 1993, pp. 16–24.

[7] Ristic, L., et al., "A Capacitive Type Accelerometer With Self-Test Feature Based on a Double-Pinned Polysilicon Structure," *Digest of Technical Papers Transducers '93*, Yokohama, June 7–10, 1993, pp. 810–813.

[8] Payne, R. S., and K. A. Dinsmore, "Surface Micromachined Accelerometer: A Technology Update," *SAE P-242 Sensors and Actuators*, 1991, pp. 127–135.

[9] Matroc, M., "An Update on Piezoelectric Ceramic Transducers," *Electronic Design*, Oct. 26, 1989, p. 119.

[10] Halvorsen, D. L., "Piezoelectric Polymer Traffic Sensors and the Smart Highway," *Sensors*, Aug. 1992, pp. 10–17.

[11] Schiller, P., D. L. Polla, and M. Ghezzo, "Surface Micromachined Piezoelectric Pressure Sensors," *Tech. Digest of IEEE Solid-State Sensor and Actuator Workshop*, Hilton Head, SC, 1990, pp. 188–190.

[12] Paranjape, M., L. Ristic, and W. Allegretto, "Simulation, Design and Fabrication of a Vertical Hall Device for Two-Dimensional Magnetic Field Sensing," *Sensors and Materials*, Vol. 5, No. 2, 1993, pp. 91–101.

[13] Chemical sensor brochure, Microsens, Neuchatel, Switzerland.

[14] Brown, C., "Single-Chip Hydrocarbon Sensor Explored," *Electronic Engineering Times*, Jan. 15, 1996, p. 41.

[15] Middelhoek, S., et al., "Sensors With Digital or Frequency Output," *Sensors and Actuators*, 15 (1988), pp. 119–133.

[16] Cumberledge, W., R. Frank, and L. Hayes, "High Resolution Position Sensor for Motion Control System," *Proc. PCIM '91*, Universal City, CA, Sept. 22–27, 1991, pp. 149–157.

[17] Ismail, M. S., and R. W. Bower, "Digital Pressure-Switch Array With Aligned Silicon Fusion Bonding," *J. Micromechanics and Microengineering*, Vol. 1, No. 4, Dec. 1991, pp. 231–236.

[18] Cigoy, D., et al., "Low Noise Cable Testing and Qualification for Sensor Applications," *Proc. Sensors Expo*, Cleveland, Sept. 20–22, 1994, pp. 575–596.

[19] Reodique, A., and W. Schultz, "Noise Considerations for Integrated Pressure Sensors," Motorola Application Note AN1646, 1998.

[20] Motorola Sensor Device Data/Handbook, DL200/D Rev. 4, 1998.

[21] Bryzek, B., J. R. Mallon, and K. Petersen, "Silicon Low Pressure Sensors Address HVAC Applications," *Sensors*, Mar. 1990, pp. 30–34.

[22] Hughes, B., "Sensing Pressure Below 0.5 psi," *Proc. Sensors Expo*, 1990, 104A1–104A5.

[23] Baum, J., "Low-Pressure Smart Sensing Solution With Serial Communications Interface," *Proc. Sensors Expo*, Boston, May 16–18, 1995, pp. 251–261.

4

Getting Sensor Information Into the MCU

Someday we'll find it—the rainbow connection.
—Kermit the Frog from Jim Henson's Muppets

4.1 Introduction

The sensor signal is only the first step toward a sensor that will ultimately provide an input to a control system. Signal conditioning the output from the sensor is as important as choosing the proper sensing technology. Most transducer elements require amplification as well as offset and full-scale output calibration and compensation for secondary parameters, such as temperature. Whether the signal conditioning is performed as an integral part of the purchased sensor or performed by the user, the accuracy of the measurement ultimately will be determined by the combination of the sensor's characteristics and the additional circuitry. The signal conditioning circuitry that is required is a function of the sensor itself and will vary considerably depending on whether the sensor is capacitive, piezoresistive, piezoelectric, et cetera.

Once a high-level analog signal is available, it must be converted to a digital format for use in a digital control system. The A/D converter (ADC) can be an integral part of the digital controller, or it may be an intermediate standalone unit. Different approaches are used for ADCs; the theoretical accuracy (resolution), indicated by the number of bits, is not always achieved in the application.

This chapter addresses signal conditioning and A/D conversion for sensors. Examples are shown of available ICs that have been designed to simplify

71

the task of signal conditioning a broad range of sensor and application circuits developed for piezoresistive pressure sensors with specific performance characteristics.

4.2 Amplification and Signal Conditioning

Micromachining is used to manufacture a diaphragm of beam thickness to nominal targets. Microelectronics is used to provide the precision for semiconductor sensors. Accuracy combined with ease of interface, cost, power consumption, printed circuit board space, and power supply voltage can be among the considerations in the selection of a signal conditioning IC. The semiconductor technologies used for the modular amplifiers and ICs in this section have a major impact on those criteria.

Sensor signal conditioning circuits are based on two fundamental technologies: bipolar and CMOS. However, those two technologies have a vast number of derivatives. The requirements of the application should be used to determine which technology or derivative is appropriate. A performance comparison between bipolar and CMOS for signal conditioning circuits is shown in Table 4.1 [1]. The comparison was performed in the early 1990s. Process technology improvements in continuously and rapidly shrinking CMOS processes have changed some of the parameters in favor of CMOS, especially for the most advanced processes. At any point in time, the two technologies can be combined to obtain BiCMOS and the best features of bipolar and CMOS at the expense of a more complex process, going from a 10-mask bipolar process to one requiring 14/15 mask steps. The BiCMOS process has lower yields because of the added processing complexity so applications must utilize the performance improvements it provides.

A 5V supply is commonly used for MCUs and DSPs. Consequently, common signal conditioned outputs for sensors are 0.5–4.5V or 0.25–4.75V. Other standard industrial outputs are 1–6V, 1–5V, and 0–6V. The supply voltage for those units can range from 7 to 30V. Data acquisition systems typically have used ±15V, ±12V, and ±5V supplies.

Signal conditioners for sensor outputs range from basic low-gain dc amplifiers to specialized amplifiers such as phase-sensitive demodulators [2]. Linear, logarithmic, high-gain, and dc-bridge amplifiers all fall in the following categories: differential amplifier, ac-coupled amplifier, chopper-stabilized amplifier, carrier amplifier, dc-bridge amplifier, and ac-level amplifier. A number of specialized amplifiers are also used for sensors, including log-linear amplifiers, frequency-to-voltage converters, integrator amplifiers, and differentiator amplifiers. Those packaged solutions take the low-level output from

Table 4.1
Comparison of CMOS and Bipolar

Circuit Element	Bipolar	CMOS
Op amp	Very good	Fair
Analog switch	Poor	Very good
Comparator	Very good	Fair
ADC	Good	Good
Reference	Very good	Poor
Mirror	Good	Very good
Regulator	Very good	Poor
Active filter	Good	Very good
Logic	Poor	Very good
Power amplifier	Good	Good
<3V operation	Very good	Good

sensors and add computational processes to free the digital controller from performing time-consuming functions. Some amplifiers work with either ac or dc excitation, and others are designed specifically for ac or dc operation.

In any sensor amplifier circuit, circulating currents in the ground path between a sensor and the measurement point can generate a common-mode voltage. The voltage appears simultaneously and in-phase at both of the differential amplifier's input terminals. The voltage can either produce an error in the measurement or a catastrophic failure. The common-mode rejection ratio (CMRR) expressed in decibels is a measure of an amplifier's ability to avoid this problem. A differential amplifier rejects common-mode voltage and amplifies only the difference across its input terminals.

4.2.1   Instrumentation Amplifiers

Several ICs have been designed to provide amplification for applications including sensors. Instrumentation amplifiers are used to provide a differential-to-single end conversion and amplify the low-level signal from a sensor. Ratiometricity, the output that varies linearly with the supply voltage, is an important consideration for sensors and signal conditioning circuitry used to amplify their output. The impedance of the sensor, the temperature range, and the desired measurement accuracy are important factors in deciding the type of signal conditioning circuitry that is required.

The instrumentation amplifier's dedicated differential-input gain block differentiates it from an ordinary op amp. Although the signal level for the sensor may be only a few millivolts, it can be superimposed on a common-mode signal of several volts. A high CMRR keeps the common-mode voltage fluctuations from causing errors in the output [3]. A simple circuit using a standard instrumentation amplifier with a gain of 10,100, CMRR of 100 dB, and bandwidth of 33 kHz is shown in Figure 4.1 [4]. The pressure sensor has passive calibration for zero offset, full-scale span, and temperature compensation, but no additional amplification. The bridge current in the circuit is less than 3.6 mA with a 9V supply, and the output can be directly interfaced to an ADC of an MCU or a DSP.

Another circuit that demonstrates the amplification and calibration for a pressure sensor is shown in Figure 4.2 [5]. This circuit has a regulated voltage for the bridge and amplifier supplied from voltage regulator U1. The supply voltage can range from 6.8 to 30V [6]. The low-cost interface amplifier provides a 0.5–4.5V output and can achieve an accuracy of ±5% by using 1%

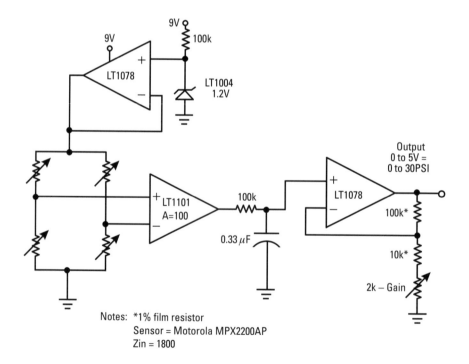

Figure 4.1 Pressure sensor circuit using an instrumentation amplifier.

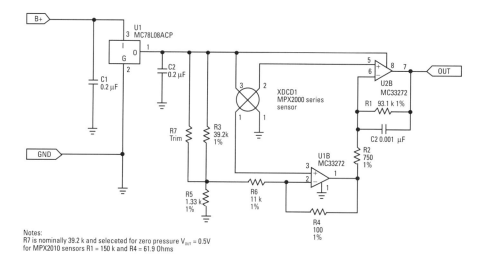

Figure 4.2 Sensor-specific signal conditioning circuit.

resistors. The MCU can increase this accuracy through software and avoid an analog trim in the amplifier.

4.2.2 SLEEPMODE™ Operational Amplifier

Integrated circuits designed to optimize performance may provide new solutions for smart sensors. For example, in battery-powered applications, the current drain and power consumption must be low to obtain optimum battery life. A bipolar operational amplifier has been designed that conserves power in portable and other low-power applications but can handle higher current drain for improved performance [7]. The SLEEPMODE operational amplifier consumes only 140 μA maximum (40 μA typical) with up to 5V when it is asleep.[1] In the awake mode, it has a typical gain bandwidth product of 2.2 MHz and a CMRR of 90 dB and can drive 700 mA. The key elements of this IC are shown in Figure 4.3. In the sleep mode, the amplifier is active and waiting for an input signal from the sensor. When a signal is applied that sinks or sources 200 μA, the amplifier automatically switches to the awake mode for higher slew rate, increased gain bandwidth, and improved drive capability. The amplifier automatically returns to sleep mode when the output current drops below the threshold (90 μA). This unit swings within 100 mV (typical) of both rails.

---

1. SLEEPMODE is a trademark of Semiconductor Components Industries, LLC (d.b.a. ON Semiconductor).

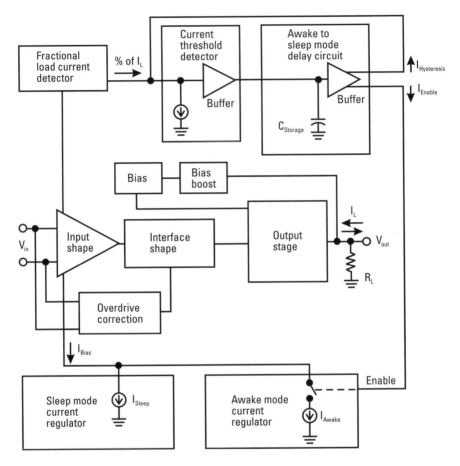

Figure 4.3 Block diagram of SLEEPMODE operational amplifier.

4.2.3 Rail-to-Rail Operational Simplifiers

Battery-powered circuits are requiring system operation at progressively lower voltages. An operational amplifier has been designed that avoids the high threshold voltage of CMOS amplifiers and operates with a single-sided supply at voltages as low as 1V or a dual supply voltage of ±0.5V [8]. The rail-to-rail design has an output stage that is current-boosted to deliver at least 10 mA to the load within 50 mV of the supply rails at 1V. The circuit can operate at voltages up to 7V. The input stage is a folded cascode that uses variable-threshold depletion mode MOSFETs. Specialized circuits of this type can be useful in providing performance for sensor signal conditioning in specific applications.

Figure 4.4 shows the effect of rail-to-rail operation [9]. Traditionally, a CMOS process has had difficulty achieving low-drift and low-noise

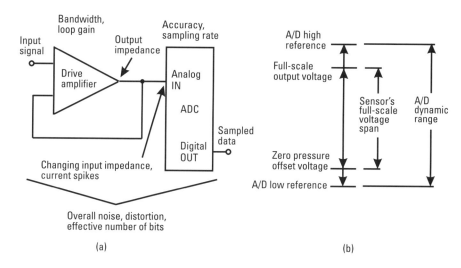

Figure 4.4 (a) Drive amplifier and ADC block diagram, and (b) effect of rail-to-rail operation on A/D performance. (*After:* [9].)

requirements of high-performance analog functions. For the amplifier to contribute no additional error to the ADC, the amplifier's signal-to-noise ratio should be better than the theoretical best-case dynamic range of the ADC. Depending on the amount of noise present, the combination of the drive amplifier and the ADC will determine the effective number of bits of resolution. The A/D dynamic range is the full-scale value (high reference minus the A/D low reference). The normal output from the sensor must be inside the A/D dynamic range to ensure proper operating headroom.

A single-pole (or multiple-pole) resistance-capacitance (RC) filter can be used to minimize the sensor noise that otherwise would be passed to an ADC [10]. In applications in which the ADC is sensitive to high source impedance, a buffer should be used. A rail-to-rail buffer amplifier is used in Figure 4.5 with an integrated pressure sensor (IPS) that also has a rail-to-rail output swing. The rail-to-rail input and output capability of the op amp avoids saturating the buffer.

4.2.4 Switched-Capacitor Amplifier

CMOS is preferred for signal conditioning sensors with capacitive output because of the high input impedance of CMOS. Another advantage of CMOS is low-power-supply current requirements. Switched-capacitor techniques

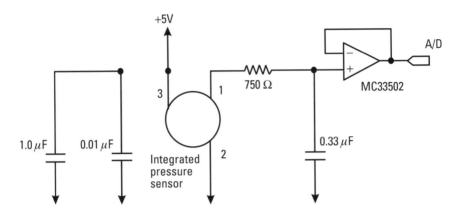

Figure 4.5 A rail-to-rail buffer used to reduce the output impedance of an RC filter [10].

using CMOS circuitry have been developed to detect changes as small as 0.5 fF. A switched capacitor filter is an analog sampled-data circuit. The components of a basic circuit are an op amp, a capacitor, and an analog switch, all driven by a clock [11]. The filter's performance is determined by the ratio of the capacitor values and not the absolute values. Switching the capacitor makes it behave as a resistor with an effective resistance that is proportional to the switching frequency. Figure 4.6 shows a switched-capacitor filter that has been developed for a surface micromachined accelerometer with differential capacitive sensing [12]. This type of circuit minimizes effects of parasitic capacitance and has a very high input impedance for minimal signal loading.

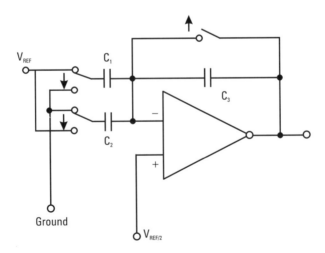

Figure 4.6 Switched-capacitor filter circuit.

4.2.5   Barometer Application Circuit

Sensor applications that use only a limited portion of the available output from a sensor can provide interesting challenges to sensor manufacturers and circuit designers. Barometric measurements, for example, use only a pressure variation of a few inches of mercury. In addition, that variation is superimposed on the altitude pressure reading, which can vary from 29.92 inches of mercury at sea level to 16.86 inches of mercury at 15,000 feet. A circuit that interfaces a temperature-compensated and calibrated 100-kPa (29.92 inches of mercury) absolute pressure sensor to a microcontroller has been designed based on a variation of Figure 4.2 [13]. The circuit has a gain of 187 and obtains a resolution of 0.1 inch of mercury when it is interfaced to an 8-bit MCU with an integral ADC.

4.2.6   4- to 20-mA Signal Transmitter

The use of monolithic two-wire transmitters can provide a low-cost, easy-to-interface signal conditioning solution for 4- to 20-mA current loops to transmit sensor readings in many industrial applications. The current loop avoids the problems of voltage drop when transmitting an analog signal over a long distance and the supply voltage, which can range from 9 to 40V. The circuitry in these devices can also compensate for nonlinearity in a resistance temperature detector (RTD) or resistance bridge sensor and therefore improve the overall performance of the sensor–signal conditioning combination. Furthermore, the availability of additional features, such as a precision current source or a 5V-shunt regulator, can simplify the task of the circuit designer and allow use of those ICs for several types of sensors. Figure 4.7 shows a high-impedance pressure sensor interfaced with a transmitter IC [14]. With zero pressure applied, the output is 4 mA; with full pressure, it is 20 mA. A 240Ω resistor referenced to ground at the receiving end provides a 0.96–4.8V signal that can be interfaced to an ADC.

4.2.7   Schmitt Trigger

One final circuit to be considered is the Schmitt trigger, which turns the pulsed output from a sensor such as an optodetector or phototransistor into a pure digital signal. Figure 4.8 shows the opto input and output of the Schmitt trigger [15]. The lower and upper thresholds in the trigger remove the linear transition region between the on and off states. This hysteresis filters electrical noise that can cause the output to change state when it is close to the threshold of a

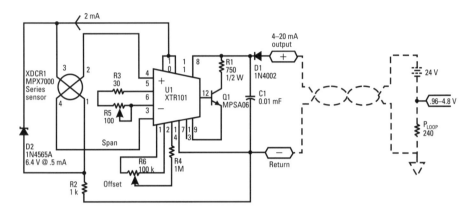

Figure 4.7 Pressure sensor with 4- to 20-mA transmitter.

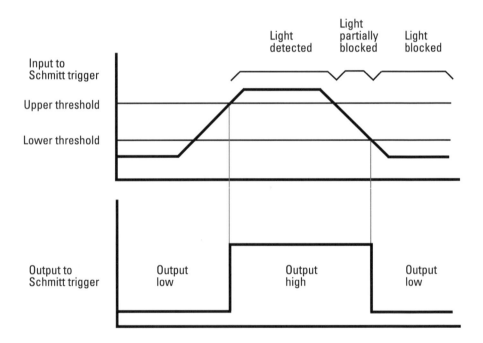

Figure 4.8 Schmitt trigger output and photodetector input.

digital IC input. The output of the Schmitt trigger can be used directly by digital logic circuits.

4.2.8 Inherent Power-Supply Rejection

If a high-voltage power supply with poor regulation is used to provide power to the sensor and the ADC input to the MCU, a software/hardware scheme can be used to reduce the errors caused by supply voltage fluctuation [16]. The scheme relies on the ratiometric properties of the sensor and the ADCs. As shown in Figure 4.9, a programmable resistor-divider array is used to determine the amount of attenuation of the unregulated 24V supply. A software algorithm relies on a calibration procedure that stores data in EEPROM. The resistor values are chosen based on the following:

- The expected range of unregulated voltages;

- The resistor array resolution necessary to attenuate the range of unregulated voltages;

- Power consumption considerations in the digital switch and programmable resistor-divider array.

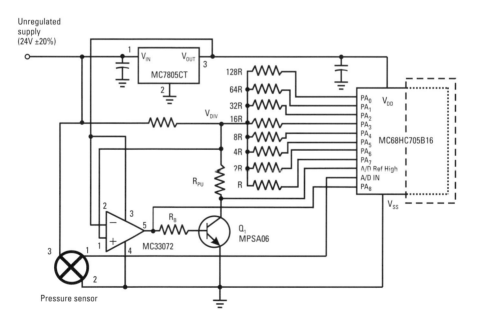

Figure 4.9 An inherent power-supply rejection circuit for a high-voltage unregulated supply.

Once the circuit is calibrated, the inherent power-supply rejection scheme operates automatically without additional software overhead or MCU processing time.

4.3 Separate Versus Integrated Signal Conditioning

There is a worldwide effort [17, 18] in both academia and industry to integrate various sensors with electronics because of the potential advantages. Those advantages include improved sensitivity, differential amplification for canceling parasitic effects such as temperature and pressure, temperature compensation circuits, multiplexing circuits, and A/D-converting circuits [19]. The combination does not inherently offer lower cost or improved performance. However, selection of the proper sensor, application (especially high-volume), and design criteria can provide a sensor that is lower cost and more reliable than a thick-film, multichip version.

Semiconductor sensors, like other semiconductor devices, are subject to the paradox that (1) increased integration is inevitable with semiconductor technology for performance and cost reduction and (2) all devices cannot be integrated and obtain improved performance or cost reduction. Proper system partitioning is required to avoid components that are difficult to integrate, such as high-value capacitors, and components that can be obtained more cost effectively by other approaches. Combining the sensor and additional circuitry can be beneficial in some applications and undesirable in others.

Precision outputs for integrated sensors can be obtained by laser trimming of zero-TCR thin-film metals at the wafer, die, or package level. Silicon resistors created by alloying silicon and aluminum can also be trimmed for improved accuracy. One trim process has achieved essentially continuous adjustment by computer-controlled current pulses applied to the resistors either at the wafer level or after packaging [20].

Integrated electronic techniques are especially useful when only a small adjustment must be made. Active elements, such as op amps and transistors, are used to produce an amplified signal that can be easily interfaced to the rest of a sensing system. Improved precision can be obtained by characterizing devices over temperature and pressure (or acceleration or force) and storing correction algorithms in memory. Semiconductor sensing technology allows the integration of both the sensing element(s) and signal conditioning circuitry into the same monolithic (silicon) structure. A number of factors determine which technology a manufacturer would choose when approaching a new design. The integration of passive elements and active elements with a piezoresistive pressure sensor is used here to demonstrate common approaches.

4.3.1 Integrated Passive Elements

A sensor with precisely trimmed offset, full-scale span, and temperature compensation can simplify the calibration procedure of the sensor and signal conditioning circuitry. Figure 4.10 shows a pressure sensor with integrated resistors trimmed to meet specifications that are possible in the semiconductor process flow. The low-TCR resistors are laser-trimmed with a 5-μm-diameter pulsed laser beam [21]. Factors that affect the trim accuracy include the variability of the resistance of the deposited thin-film metal resistors, the annealing cycle for stress relief, and the sheet resistance. Process control allows the trimmed product to be extremely repeatable and provides high volume and high yield. That small amount of integration simplifies subsequent amplification circuitry, as the circuits in Figure 4.1 and Figure 4.2 demonstrated. It also allows the sensor to be applied to a number of applications and with varied supply voltages.

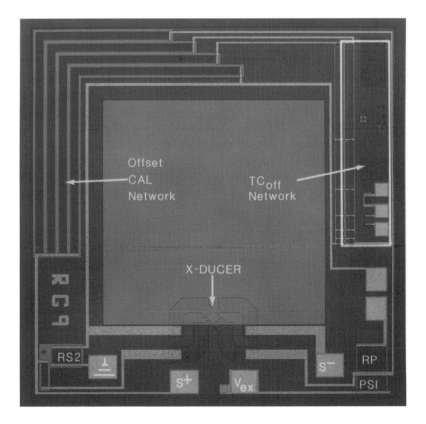

Figure 4.10 Photomicrograph of micromachined piezoresistive pressure sensor with integrated resistor network for zero offset, span, and temperature compensation.

4.3.2 Integrated Active Elements

Integrating the amplification and signal conditioning directly on the sensor can be accomplished by the use of combined micromachining and micro-electronics. The circuit of a fully signal conditioned pressure sensor is shown in Figure 4.11(a). The additional circuitry is integrated on the sensor using the silicon area that is required to provide the mechanical support for the dia-phragm [22]. The die size for the fully signal conditioned unit is 0.145 inch by 0.130 inch. The diaphragm and piezoresistive element account for 20% of the die, and the signal conditioning circuitry is 80%, including the area required for wire bond pads. A quad operational amplifier frequently used for sensor output amplification is the LM324. The area of silicon in the LM324 is about 0.050 inch by 0.050 inch. The sensor without signal conditioning is a 0.120-inch by 0.120-inch chip. In that case, the electronics has been combined with the sensor with a very small increase in the total chip area because the mechani-cal support for the diaphragm has been utilized. Common pressure ranges and a standard interface to 5V dc ADC allow the integrated approach to satisfy a number of application requirements and justify the cost of the integrated design. The integrated sensor easily interfaces to an MCU with an onboard ADC, as shown in Figure 4.11(b). The 50-pF capacitor and 51-kΩ resistor are recommended as a decoupling filter.

A more recent circuit design met the target specifications with one less stage in the amplification circuitry than the earlier design [23]. As shown in Figure 4.12, it uses fewer resistors and is easier to trim. In addition a higher source/sink current can be obtained in the new design. Initial indications are that improvement of at least 10 times has been obtained. The output stage has been designed to drive CMOS A/D inputs from 0.2V to 4.7V (typical values) with a 5V supply.

Along with the new circuit, a revised trim algorithm was developed. Although the basic principles are similar to the earlier version, there are several important improvements. One of the most important is the balanced design, which allows for bidirectional trimming of all parameters, and a more inde-pendent trim of each parameter. On the old circuit, parameters such as tem-perature coefficient of offset (Tco) and gain could be trimmed in only one direction. If processing variations resulted in the initial value of those parame-ters being beyond the target values, the device could not be trimmed. Addition-ally, many parameters had to be trimmed with an iterative process in which several parameters were roughly trimmed, and then they were all trimmed again, to reach the final target. Any adjustment of one resistor affected more than one parameter and limited the overall system accuracy. For the new design, all parameters can be trimmed in either direction, and each one is

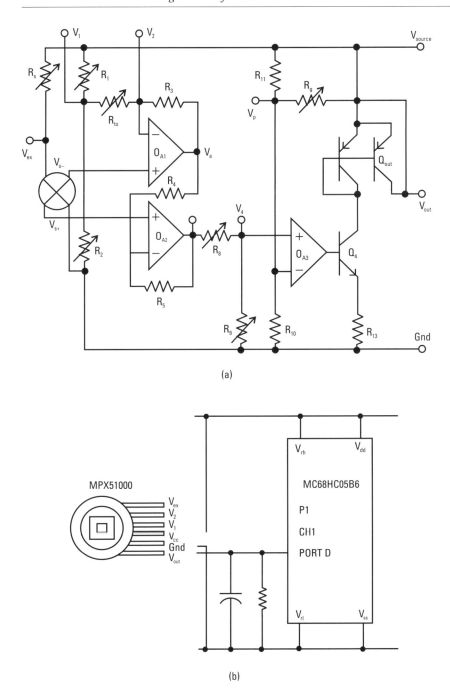

(a)

(b)

Figure 4.11 (a) Integrated piezoresistive pressure sensor circuit and interface to MCU, and (b) sensor interfaced to MCU.

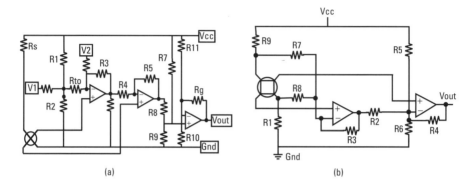

Figure 4.12 Circuit comparison: (a) original 3–op amp design, and (b) reduced component count in the new 2–op amp design.

measured and trimmed independently of the others. That results in more accurate trim capability with tighter manufacturing control.

The new device has an on-chip heater that is controlled independently from the rest of the circuit. When the Tco trim is reached, the heater is turned on and a controlled voltage waveform is applied, so that the die is heated and rapidly reaches a steady-state temperature. The Tco is measured and compensated, and then the heater is deactivated. The new method achieves a more uniform temperature profile, as well as reduced thermal response time and total Tco trim time. That level of improvement from one design level to the next is part of the smarter approach that microelectronics can and must provide to sensing.

4.4 Digital Conversion

Various A/D architectures are available for integration and interfacing with sensors. Conversion resolution, conversion accuracy, conversion speed or bandwidth, inherent system noise levels, and power consumption are all ADC tradeoffs. In assessing a converter architecture, it is important to consider all those aspects. For example, errors due to temperature, supply voltage, linearity, quantizing, and so on, may reduce the accuracy of an ADC by several bits when all error sources are considered. Also, bit accuracy alone may not be sufficient, especially if the sampling or conversion rate is incorrect for the sensor response under consideration [24]. Table 4.2 shows the quantizing errors and other parameters for 4- to 16-bit A/D conversion [25]. The quantization error (as a percentage of full-scale range) is $\pm 1/2 \cdot 1/(2^n - 1) \cdot 100$, which is also $\pm 1/2$ LSB. The resolution is the least significant bit percentage of full scale/100. The

Table 4.2
A/D Bits and Dynamic Range

A/D (Number of Bits)	LSB Weight % of FS	LSB Voltage for 5V FS	Dynamic Range (dB)
4	6.25	300 mV	24.08
8	0.3906	19.5 mV	48.16
10	0.0977	4.90 mV	60.12
12	0.0244	1.20 mV	72.25
14	0.00610	305 μV	84.29
16	0.00153	75 μV	96.33

theoretical root mean square (rms) signal-to-noise ratio for an N-bit ADC is calculated by

$$\text{signal-to-noise ratio} = 6.02 \cdot N + 1.76 \, \text{dB} \qquad (4.1)$$

where N = number of bits.

4.4.1 A/D Converters

Common A/D conversion techniques include single slope (ramp-integrating), dual slope integrating, tracking, successive approximation, folding (flash), and sigma-delta oversampled ADCs. Table 4.3 compares six types and lists the relative conversion rate and relative silicon area required for each [24]. Note the effect of hardware- versus software-driven successive approximation.

Most ADCs can be classified into two groups based on the sampling rate, namely, Nyquist rate and oversampling converters. The Nyquist rate requires sampling the analog signals that have maximum frequencies slightly less than the Nyquist frequency, $f_N = f_s/2$, where f_s is the sampling frequency. However, input signals above the Nyquist frequency cannot be properly converted and create signal distortion or aliasing. A low-pass antialiasing filter attenuates frequencies above the Nyquist frequency and keeps the response below the noise floor.

Sigma-delta (Σ-Δ), or delta-sigma, converters are based on digital filtering techniques and can easily be integrated with DSP ICs. Σ-Δ converters sample at a frequency much higher than the Nyquist frequency. Figure 4.13 shows the

Table 4.3
ADC Architectures

ADC Type	Typical Number of Bits	Relative Conversion Rate	Relative Die Area
Folding (flash)	8	Fastest	14
Successive approximation (hardware driven)	12	Fast	10
Sigma-delta (Σ-Δ)	16	Slow	8
Successive approximation (software driven)	12	Slow	7
Dual slope	12	Slowest	3
Single slope	12	Slowest	1

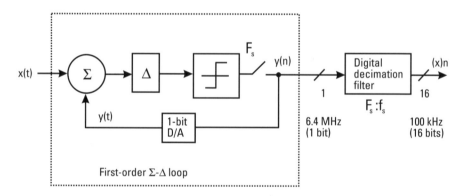

Figure 4.13 Block diagram of a first-order Σ-Δ ADC.

block diagram of an oversampled first-order Σ-Δ ADC [26]. The analog input is summed at the input node with the difference of the output of the 1-bit digital-to-analog converter (DAC). The resulting signal is provided to an integrator and then to a 1-bit quantizer (ADC). A second-order Σ-Δ consists of two integrators, two summers, and the 1-bit quantizer. Second- or third-order Σ-Δ modulators reduce the baseband noise level even further than a first-order Σ-Δ unit.

The output of the Σ-Δ converter is averaged by applying it to the input of a digital decimation filter. The digital decimation filter performs three functions. First, it removes out-of-band quantization noise, which is equivalent to increasing the effective resolution of the digital output. Second, it performs decimation, or sample rate reduction, bringing the sampling rate down to the Nyquist rate. That minimizes the amount of information for subsequent transmission, storage, or signal processing. Finally, it provides additional antialiasing rejection for the input signal.

4.4.2 Performance of A/D Converters

Frequently the sensor requires high resolution at moderately fast conversion rates. In such cases, signal gain ranging and/or offset zeroing with 8-bit ADCs can be used in a cost-effective manner. Input ranging/zeroing allows for an increase in dynamic range without compromising the total conversion time or significantly increasing the silicon area and hence cost. As shown in Figure 4.14, 10 bits of signal input can be achieved by range amplification and/or offset zeroing prior to an 8-bit ADC. Because the ranging/zeroing is typically defined by the settling time of the operational amplifier, the total conversion time remains small in comparison to a full 10-bit A/D conversion. Eight bits of accuracy is not a limitation depending on the application, because many processes normally operate around a rather small portion of the full-scale signal. A high degree of accuracy is required in the operating range, especially if the sensor is being used as a feedback element. By adjusting the input gain and offset, an 8-bit ADC can be placed at the optimum measurement point but still be able to react to sudden excursions outside the normal range [24].

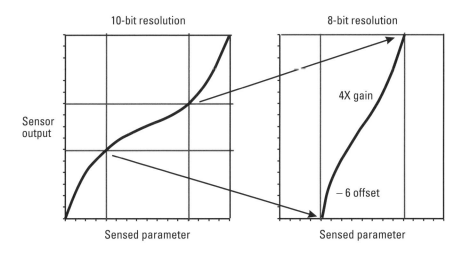

Figure 4.14 Increasing the accuracy of an 8-bit ADC.

ADC resolution can also be increased with circuit techniques. For example, a circuit that uses both ADC and DAC channels of an MCU and a two-stage amplifier has been designed that increases the effective resolution of an 8-bit ADC to 12 bits for pressure measurements [27]. Standard components were used for the MCU and amplifier circuit. An applied pressure in the mid-range of a 200-kPa sensor originally measured with an error of 0.3 kPa was subsequently reduced to 0.1 kPa using that approach.

Another consideration in choosing an ADC is specification of the requirements within the noise sources of the system. Those limitations may be due to a less-than-optimum printed circuit layout, unstable power sources, nearby high-energy fields, and/or use of devices without good signal-to-noise ration and power-supply CMRR. It is important to first assess the noise floor for the system and then compare that to the desired resolution of the signal. If the combined system's design and noise floor conditions create a minimum required signal that must be greater than 5–10 mV, only 8–9 bits of resolution can be determined in a 5V system. Using a higher resolution ADC will not provide any additional data unless the conversion speed of the ADC and processor throughput allows multiple sampling with data correction. Furthermore, the process must also meet the desired bandwidth, power consumption, and system response [28]. Oversampling and averaging are, therefore, desired because it is not possible to separate the noise from the signal using a single A/D sample/conversion.

4.4.3 Implications of A/D Accuracy and Errors

An amplified pressure signal supplied to an 8-bit ADC is an example of the combined capability of the amplifier and the ADC. The resulting A/D conversion is related to the pressure input by the following equation:

$$\text{count} = \left(V_{FS} - V_{Offset}\right) \cdot 255 / \left(V_{RH} - V_{RL}\right) \tag{4.2}$$

Where $V_{FS} - V_{Offset}$ is the sensor's full-scale span voltage, 255 is the maximum number of counts from the 8-bit converter, and $V_{RH} - V_{RL}$ is 5V, based on using the same 5V supply as the MCU. Using the same reference voltage for the ADC and the sensor minimizes the number of additional components but does sacrifice resolution. For those instances where greater resolution is required, separate lower voltage references should be provided for the ADC and/or the sensor.

For a sensor with a 0.25–4.75V output, the maximum number of counts available at the output register will be *count* (full scale) = 229

A full-scale pressure of 15 psi with 5.0V supply results in a system resolution of

15 psi /229 = 0.066 psi /*count*

4.5 Summary

Several circuits and techniques can be used to amplify the low-level output that is inherently available from sensor or transducer elements. The circuits that were discussed are used primarily for interfacing the analog output of the sensor to the digital world of the microcontroller and the DSP. The ongoing industry effort to improve performance and simplify the interface for sensors either as standalone units or integrated with the sensors will continue to expand the applications for sensors in general and simplify the design of future smart sensors.

References

[1] Dunn, B., and R. Frank, "Guidelines for Choosing a Smart Power Technology," *PCI '88*, Munich, June 6–8, 1988.

[2] Klier, G., "Signal Conditioners: A Brief Outline," *Sensors*, Jan. 1990, pp. 44–48.

[3] Conner, D., "Monolithic Instrumentation Amplifiers," *EDN*, Mar. 14, 1991, pp. 82–88.

[4] Williams, J., "Good Bridge-Circuit Design Satisfies Gain and Balance Criteria," *EDN*, Oct. 25, 1990, pp. 161–174.

[5] Schultz, W., "Amplifiers for Semiconductor Pressure Sensors," *Proc. Sensors Expo West*, San Jose, CA, Mar. 2–4, 1993, pp. 291–298.

[6] Motorola Application Note AN1324, *Sensor Device Data/Handbook*, DL200/D Rev. 4, 1998.

[7] Motorola Data Sheet MC33304/D, 1996.

[8] McGoldrick, P., "Op Amp Supply Squeezed Down to 1V Rail-To-Rail," *Electronic Design*, Feb. 17, 1997, p. 87.

[9] Motorola Application Note AN1646, 1998, Phoenix, *Sensor Device Data/Handbook*, DL200/D Rev. 4, 1998.

[10] Swager, A. W., "Evolving ADCs Demand More From Amplifiers," *Electronic Design News*, Sept. 29, 1994, pp. 53–62.

[11] Baher, H., "Microelectronic Switched Capacitor Filters," *IEEE Circuits and Devices*, Jan. 1991, pp. 33–36.

[12] Dunn, W., and R. Frank, "Automotive Silicon Sensor Integration," SAE SP-903 *Sensors and Actuators 1992*, Detroit, Feb. 24–28, 1992.

[13] Winkler, C., and J. Baum, "Barometric Pressure Measurement Using Semiconductor Pressure Sensors," *Proc. Sensors Expo West*, San Jose, CA, Mar. 2–4, 1993, pp. 271–283.

[14] Motorola Application Note AN1318, *Sensor Device Data/Handbook*, DL200/D Rev. 4, 1998.

[15] Cianci, S., "Airborne Optical Encoders," *EDN Products Edition*, Jan. 16, 1995, pp. 9–10.

[16] Jacobsen, E., "Inherent Power-Supply-Rejection Scheme Improves Sensor-System Performance," *EDN*, Nov. 7, 1996, pp. 161–164.

[17] Kandler, M., et al., "Smart CMOS Pressure Sensor," *Proc. 22nd Internat'l Symp. Automotive Technology Automation* #90185, Florence, Italy, May 14–18, 1990, pp. 445–449.

[18] Iida, M., et al., "Electrical Adjustments for Intelligent Sensor," *SAE 900491*, Society of Automotive Engineers Internat'l Congress and Exposition, Detroit, Feb. 26–Mar. 2, 1990.

[19] Senturia, S. D., "Microsensors vs. ICs: A Study in Contrasts," *IEEE Circuits and Devices*, Nov. 1990, pp. 20–27.

[20] Frank, R., and J. Staller, "The Merging of Micromachining and Microelectronics," *Proc. 3rd Internat'l Forum on ASIC and Transducer Technology*, Banff, Alberta, Canada, May 20–23, 1990, pp. 53–60.

[21] Staller, J. S., and W. S. Cumberledge, "An Integrated On-chip Pressure Sensor for Accurate Control Applications," *IEEE Solid-State Sensors and Actuators Workshop*, Hilton Head Island, SC, 1986.

[22] Frank, R., "Two-Chip Approach to Smart Sensors," *Proc. Sensors Expo*, Chicago, 1990, pp. 104C1–104C8.

[23] Bitko, G., R. Frank, and A. McNeil, "Improvements From Next-Generation MEMs Sensor Design" (to be published).

[24] Frank, R., J. Jandu, and M. Shaw, "An Update on Advanced Semiconductor Technologies for Integrated Smart Sensors," *Proc. Sensors Expo West*, Anaheim, CA, Feb. 8–10, 1994, pp. 249–259.

[25] Hoeschele, D. F., Jr., *Analog-to-Digital and Digital-to-Analog Conversion Techniques*, New York: Wiley, 1993.

[26] Park, S., "Principles of Sigma-Delta Modulation for Analog-to-Digital Converters," Motorola APR8/D Rev. 1, 1993.

[27] Motorola Application Note AN1100, *Sensor Device Data/Handbook*, DL200/D Rev. 4, 1998.

[28] Johnson, R. N., "Signal Conditioning for Digital Systems," *Proc. Sensors Expo*, Philadelphia, Oct. 26–28, 1993, pp. 53–62.

5

Using MCUs/DSPs to Increase Sensor IQ

We could use cow chips instead of microchips and save millions.
—Dilbert™, by Scott Adams

5.1 Introduction

Several semiconductor technologies are available to improve the accuracy and the quality of the measurements and to add diagnostics and other intelligence to any type of sensor. Foremost among those technologies are MCUs, DSPs, application-specific ICs (ASICs), and field-programmable gate arrays (FPGAs). Dedicated sensor signal processors are usually an adaptation of one of those approaches. The technologies also have the potential to allow for a fully integrated (monolithic) smart sensor. Before taking this next, rather large step, it is important to understand the technologies that are available, their contribution to smart sensors, and their ability to provide a higher level of intelligence (and value) to sensors.

5.1.1 Other IC Technologies

ASIC technology utilizes computer-aided design (CAD) software tools to achieve custom circuit designs. ASIC technology consists of programmable logic devices (PLDs) for low-circuit density (<5,000 gates), gate arrays for medium density (<100,000 gates), and standard cells for high-end custom circuits. ASIC devices combine high density and integration of full custom designs with relatively low cost and fast design turnaround. A custom highly integrated chip utilizing "core" microprocessor cells combined with analog, memory, and additional logic functions can address specific sensing requirements such as fluid-level sensing. Mixed-signal ASICs combine analog with digital capability.

FPGAs and field-programmable analog arrays (FPAA), analog versions of FPGAs, are attractive as sensor interfaces because they can minimize development time and can be reconfigured after they have been in service. Both the FPGA and FPAA use a front-end circuit design program [1]. A circuit design is transferred to the FPGA (or FPAA) by either downloading converted serial data directly to the on-chip static random access memory (RAM) in the FPGA or to serial erasable programmable ROM (EPROM). A digital core incorporated in the FPGA design gives it the same kind of computing capability as an MCU with the advantages of field programmability. FPGAs are especially useful when rapid circuit prototyping and flexibility are required.

The term *system on a chip* (SOC) is frequently used to describe a highly integrated circuit, ASIC, MCU, or DSP that incorporates considerably more hardware options than previously available versions. These chips frequently incorporate application-specific software that is closely linked to the hardware on the chip. These chips include a large and increasing amount of the system and reduce the total chip count. In only the simpler systems are they actually the whole system. That is especially true with the sensor portion of the system. The added complexity of adding the sensor to the SOC makes a sensor system on a chip far more difficult—and much more expensive—than available alternatives. (Chapter 14 provides more detail on the sensor SOC.)

5.1.2 Logic Requirements

The shift of the logic requirements from a centralized computer to nodes in decentralized systems is creating the need for smart sensors. Sensor-driven process control systems that eliminate human operators and increase the precision of the process will play an important role in the manufacturing of the semiconductors that control them. A new architecture and interface have been proposed by researchers at the University of Michigan to address this application [2]. Figure 5.1 shows the block diagram of the key elements of that proposal. Amplification and A/D conversion were discussed in Chapter 4, and the communications interface is covered in Chapter 7. This chapter uses existing MCU and DSP products to demonstrate the remaining elements. Other system components that can be obtained with an MCU or a DSP are also discussed.

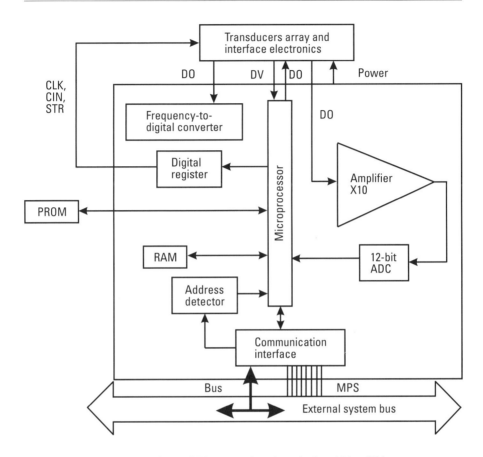

Figure 5.1 Block diagram of monolithic sensor interface design. (*After:* [2].)

5.2 MCU Control

Single-chip MCUs combine microprocessor unit (MPU) computing capability, various forms of memory, a clock oscillator, and I/O capability on a single chip, as shown in Figure 5.2 [3]. MCUs provide flexibility and quick time-to-market for numerous embedded control systems and for smarter sensing solutions. The programmability and wide variety of peripheral options available in the MCU provide design options. Those options can offset the cost of the additional component by eliminating other components or providing features that would otherwise require far more components. In addition, those high-volume products enable systems to achieve low cost, high quality, and excellent reliability.

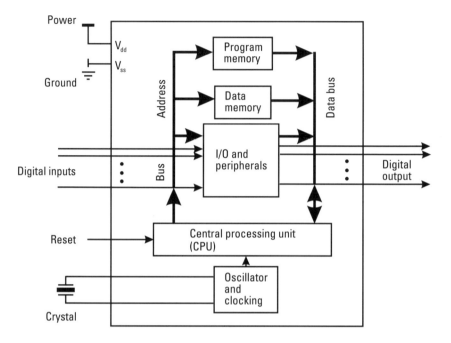

Figure 5.2 Basic MCU block diagram.

5.3 MCUs for Sensor Interface

In addition to the basic features of an MCU, a number of custom modules are integrated on the same chip to increase the utilization of the process, reduce printed circuit board space, and increase the functionality for a specific application. This section discusses MCU features that have a significant effect on sensor system performance. These include analog input capabilities, A/D conversion techniques, processing bandwidth, electronically programmable trim, onboard memory, power conservation, and improved electromagnetic compatibility (EMC) and control of radio frequency interference (RFI). For purposes of discussion here, the MC68HC05 family of 8-bit MCUs is used to explain the various MCU features. The same features, possibly with different specifications, perform similarly on other families of MCUs and higher performance 16-bit and 32-bit products as well.

5.3.1 Peripherals

Existing MCU technology contains a variety of peripherals, or hardware options, that enhance the capability of the MCU. Peripherals enable the MCU to obtain information from sensors and control output devices. Some of the most common peripherals are general-purpose I/O ports, timers, and serial ports.

Timers usually measure time relative to the internal clock on the chip or an externally provided clock signal. An on-chip oscillator that operates up to 4.0 MHz at 5V or 1 MHz at 3V controls the clock on the chip. A more complex timer can generate one or more PWM signals, measure the pulse width, and generate additional output pulse trains.

Two basic serial ports are the serial communications interface (SCI) and the serial peripheral interface (SPI). The SCI is a simple two-pin interface that operates asynchronously. Data is transmitted from one pin and received on the other. Start and stop bits synchronize communications between two devices. The SCI port is a universal asynchronous receiver transmitter (UART) that can be used with an RS-232 level translator to communicate with personal or other types of computers over fairly long distances.

The SPI port requires a third pin to provide the synchronizing signal between the control chip and an external peripheral. This type of communication is usually on the same board. Standard SPI peripherals are available from many manufacturers and include ADCs, display drivers, EEPROM, and shift registers.

5.3.2 Memory

Various types of memory can be integrated on a chip, including RAM, read only memory (ROM), EPROM, EEPROM, and flash memory. Semiconductor memory is based on a single transistor or cell that is on or off to generate a bit that is either a 1 or a 0. Memory is classified as either volatile or nonvolatile. Volatile memory is not stored when the power is disconnected to the MCU. Nonvolatile (NV) memory is stored when power is disconnected. The amount of memory in a chip is usually rated in kilobytes (1 KB = 1,024 bits). Increasing the amount of memory increases the chip size and the chip cost. Some types of memory, such as EEPROM, can significantly increase the process complexity and also add to the cost.

RAM can be read or written (changed) by the CPU and is volatile [3]. The write/read (W/R) endurance is a key parameter for RAM. Dynamic RAM requires a frequent refresh signal or the data will be lost. Static RAM does not require a refresh signal. ROM, which can be read but not changed, is nonvolatile memory and is included in the design (masked layout) of the chip. Reprogramming a chip once it has been designed is a common practice to correct errors in the original software, to upgrade to improve system performance, or to adjust for variation that could have occurred since the system was initially installed. EPROM can be changed by erasing the contents with an ultraviolet light and then reprogramming new values. That nonvolatile memory has a limited number of erasure and reprogramming operations. One-time programmable (OTP) ROM is the same as EPROM except that it is packaged in a lower cost opaque package. Because ultraviolet light cannot penetrate the package, this memory cannot be erased after it is programmed. EEPROM is a nonvolatile memory that can be changed by using electrical signals. Typically an EEPROM location can be erased and reprogrammed thousands of times before it wears out. One of the newest types of memory is flash memory. Flash memory is nonvolatile memory that is easily reprogrammed in the application faster than EEPROM. Once it is programmed, flash memory contents remain intact until software initiates an erase cycle. Program and erase voltages for EEPROM and flash memory are performed at approximately 12V. A next-generation memory still in development is ferroelectric random access memory (FeRAM or FRAM). The *ferroelectric effect* is the tendency of dipoles within a crystal to align in the presence of an electric field and to remain polarized after the field is removed [4]. Reversing the field causes polarization in the opposite direction. No current is required to maintain either state providing a binary memory capacitor with low power consumption. FeRAM for sensing applications has several advantages:

• Very fast write times (up to 20 times faster than EEPROM);
• Write/erase (W/E) endurance up to 10 million times greater than EEPROM;
• Arrays up to eight times larger than prior versions;
• Lower voltage, lower power operation to conserve battery life.

Table 5.1 is a comparison of five different types of memory [4].

5.3.3 Input/Output

I/O is a special type of memory that senses or changes based on external digital elements and not the CPU [3]. I/O ports connect the external elements to the CPU and provide control capability for the system. I/O can be either parallel, transferring 8 bits at a time to the MCU, or serial, transferring data 1 bit at a time.

General purpose I/O connections (pins) can be used as either an input or an output. A number of pins are typically grouped together and called a port. The program determines the function of each pin. Program instructions

Table 5.1
Comparison of Semiconductor Memories (*After:* [4])

	FeRAM	DRAM	Static RAM (SRAM)	Flash	EEPROM
Cell density*	3.5	2	5	1	4.5
Nonvolatility	Yes	No	No	Yes	Yes
Rewrite speed	150~200 ns	50~100 ns	5~100 ns	10 μs~1 ms	1 ms
W/R endurance	(W/R) 10^{10}~10^{13}	Infinite	Infinite	(W/E) $< 10^6$	(W/E) $< 10^5$
Power dissipation	Very low	Medium	Medium	Low	Medium
Feature in application	Battery backup equipment	High-density memory	High-speed memory	File and card memory	Low-density NVROM

*1 = smallest; 5 = largest.

evaluate the logic state of each input and drive outputs to logic 1 or 0 to implement the control strategy. Input-capture and output-compare functions in the MCU simplify the design of the control strategy.

Input-capture is used to record the time that an event occurred. By recording the time for successive edges on an input signal, software can determine the period and the pulse width of the signal. Two successive edges of the same polarity are captured to measure a period. Two alternate polarity edges are captured to measure a pulse width [5].

Output-compare is used to program an action at a specified time. For example, an output is generated when the output-compare register matches the value of a 16-bit counter. Specific duration pulses and time-delay are easy to implement with this function.

5.3.4 Onboard A/D Conversion

Various types of ADCs were discussed in Chapter 4. An ADC is frequently integrated with the MCU. For example, the typical ADC in the HC05 family of MCUs consists of an 8-bit successive approximation converter and an input-channel multiplexer. Some of the channels are available for input, and the others may be dedicated to internal test functions. For example, the 8-bit ADC of the MC68HC05B6 has a resolution of about 0.39%. The reference supply for the converter uses dedicated input pins to avoid voltage drops that would occur by loading the power supply lines and subsequently degrading the accuracy of the A/D conversions. To achieve ratiometric conversion, the +5V supply to the sensor is also connected to the V_{RH} reference input pin of the ADC, and the ground is referenced to V_{RL}. An 8-bit status control register is used to indicate that the oscillator and current sources have stabilized and that the conversion has been completed. The results of the conversion are stored in a dedicated 8-bit register.

For those instances when higher resolution is required, the SPI port of the 68HC05 allows external circuitry to be interfaced. For example, an integrated circuit such as Linear Technologies' LTC1290 connected to the SPI clock, data in, data out, and one additional programmable output pin provides a four-wire interface for a 12-bit data conversion. The data are transferred in two 8-bit shifts to the 68HC05 in 40 μs. By adding the 12-bit capability, the resolution is improved from 0.39% to 0.0244%.

The successive approximation register (SAR) is the most popular method of performing A/D conversions because of its fast conversion speed and ease of use with multiplexed input signals. The 8-bit SAR ADC on the MC68HC05P8 MCU has the timing shown in Figure 5.3 and can be driven from the processor bus clock or an internal RC oscillator running at approximately 1.5 MHz [6]. That operating frequency makes the overall time to access and convert one signal source approximately 16–32 μs. Using the Nyquist criteria for 2X sampling per cycle, such an A/D rate could be used for input sources up to approximately 31 kHz, which is well above that needed for many pressure, temperature, and acceleration sensors. With that increased bandwidth, additional samples can be taken and averaged to reduce effects of random noise sources and aliasing from higher frequency components. A simple averaging of four consecutive samples with this ADC can take less than 128 μs for an effective bandwidth of approximately 3.9 kHz. Since this ADC also has a

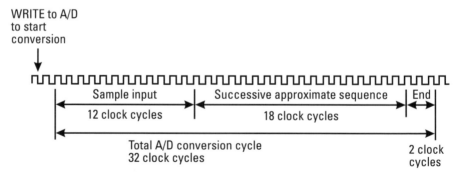

Figure 5.3 A/D process timing considerations for MC68HC05P8.

hardware-driven SAR, each conversion can be started and performed without decreasing CPU processing power. Meanwhile, the CPU can process prior A/D samples until it is interrupted with a new conversion result. Some variations of this ADC on the higher performance 68HC11 family have hardware capable of automatically storing four consecutive readings into four separate results registers.

Another A/D consideration is the use of a sample-and-hold of the signal before the SAR operation begins. The ADC in the HC05 family performs that function in typically 6–12 μs. If the input is sampled too long, the signal must not change more than one-half LSB during sampling to limit errors. That maximum change of one-half LSB in the input signal during the sample time can severely limit input bandwidth considerably more than the minimum 2X sampling Nyquist criteria.

A further advantage of an ADC design with an internal RC oscillator is the ability to optimize power versus bandwidth. That is accomplished by providing a short A/D conversion cycle while the main processor is run at a much lower bus frequency. This power-saving approach combined with others can increase useful battery life in portable applications.

5.3.5 Power-Saving Capability

Another advantage of MCU hardware and software is the variety of power-saving approaches. Varying the processing speed or stopping processing altogether can have a significant effect on overall power consumption. In addition, the ability to operate at lower voltages also reduces power consumption. Reducing power consumption involves more than reducing the supply current while running processor code.

The 68HC05 MCU family has two additional modes of operation to reduce power consumption, called the WAIT and STOP modes. In the WAIT mode, the processor bus is halted but the on-chip oscillator and internal timers are left in operation. The device returns to normal operation following any interrupt or an external reset. The WAIT mode in the MC68HC05P8 device reduces current consumption by nearly a factor of 2 at any given processor speed. In the STOP mode, both the processor and the on-chip oscillator are halted. The device returns to normal operation only after an external interrupt or reset. The STOP mode in the MC68HC05P8 reduces supply current consumption to less than 180 μA with a 5V dc supply operating at temperatures below 85°C.

An example demonstrates the importance of modes of operation and ADC performance versus power consumption of the signal processor. Consider a signal processing example using a procedure that requires 100 CPU instructions following an A/D conversion of a pressure sensor with a 500-Hz bandwidth. The MC68HC05P8 modes of operation allow several approaches, as shown in Figure 5.4 [7]. Assuming a typical CPU cycle of 1 μs (1-MHz bus), the MC68HC05P8 will require 32 μs for an A/D conversion and an additional 300 μs for the digital procedure (approximately three cycles per instruction). A total system process time is 332 μs and would produce a 6X sampling of a 500-Hz signal.

If the MC68HC05P8 in Figure 5.4 is operated at a 1-MHz bus speed at 5V dc over the temperature range of −40°C to +85°C, continuous conversions and processing will typically consume an average of 2.38 mA of supply current. However, if the device is only allowed to sample, convert, and process at two times the Nyquist criteria of once every millisecond, it could sample and process in 332 μs and then go into the WAIT mode. The on-chip timer would wake it up each millisecond. The resulting supply current would drop to an average of about 1.32 mA, as shown in Figure 5.4. Another alternative to save power would slow the system clock to process a sample every millisecond. In that case, the CPU clock could be dropped to 3 μs (333-kHz bus) while the supply current to the MC68HC05P8 would drop to about 0.79 mA, as shown in Figure 5.4. Other combinations of bus speed and WAIT times would require average supply currents from 0.79 to 1.32 mA.

The STOP mode is useful in cases in which the oscillator's restart time of typically 4064 CPU cycles is not a significant portion of the sampling/process

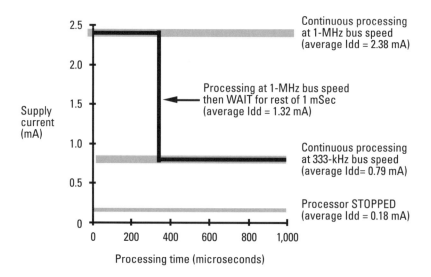

Figure 5.4 Processing mode effects on power consumption.

time and data acquisition, and processing is requested by an external source. For example, that occurs if a serial signal is transmitted every second following a temperature reading. Using the STOP mode, the supply current for the previous example would drop from about 2.38 mA for continuous 1-MHz bus operation to an average current near 0.18 mA. The sample, convert, process, and send sequence occurs at 1 MHz, then stops and waits for another external request to restart the sequence almost a second later.

5.3.6 Local Voltage or Current Regulation

Onboard voltage or current regulation is important to sensors that are not ratiometric, because the variation in supply voltage over a –40°C to +125°C operating range can be greater than ±5%. The availability of analog control circuitry with 40V standoff capability allows the integration of a series pass 5V regulator on the MCU (e.g., the MC68HC705V8) [7]. Availability of a similar shunt regulator allows a two-wire self-protected and self-powered system to be designed using only a sensor and an MCU. At higher levels of integration, such analog voltage and/or current regulation can both reduce component count and improve accuracy. Furthermore, those regulation schemes can be dynamically altered to improve functionality or reduce power consumption.

5.3.7 Modular MCU Design

A methodology has been developed that allows custom MCUs to be designed to specifically address the requirements of a particular application [7]. The customer-specified integrated circuit (CSIC) approach differs from an ASIC in the performance and density that can be achieved. Typical ASIC chip solutions utilize (1) a family of basic elements designed to handle a variety of requirements and (2) automated design tools to reduce the design cycle time. Both of those attributes contribute to a larger die size. The automated chip assembly methods may also compromise analog performance when low-level signals are involved.

CSICs utilize standard functional subsystems that have a proven field history in applications, including the demanding environment of automobiles. The chassis consists of pretested reusable blocks based on existing 68HC05 MCUs. Those modular blocks include RAM, ROM, EPROM, and EEPROM from existing modules as well as serial communications modules, a variety of display drivers such as liquid crystal display (LCD), vacuum fluorescent display (VFD), or light-emitting diodes (LED), timers, and A/D and D/A converters (Figure 5.5). These modules can be modified, or existing modules from previously developed products can be utilized. Also, new modules can be designed to

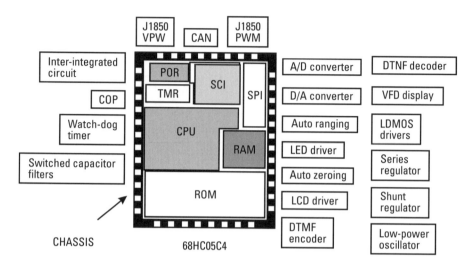

Figure 5.5 Custom MCU and various building blocks.

meet a specific customer requirement. The last case, "design to order," is used for only a small portion of the total design. The preferred approach utilizes hand-packed, highly optimized, and field-tested circuits to achieve short design cycle time, first-time success, and low end-product cost.

The CSIC approach minimizes—in some cases eliminates—nonrecurring engineering costs, and it allows a custom chip to be designed that can cost less than $1 for some units. In many cases, one of the existing designs is acceptable to provide a "semicustom" solution without requiring a new design. The number of available MCU options and a methodology that allows quick customization of MCUs have been in place for several years to address the specific requirements of various applications.

5.4 DSP Control

DSPs have hardware arithmetic capability that allows the real-time execution of feedback filter algorithms. In contrast, MCUs use lookup tables to approximate filter algorithms with inherent limitations of flexibility and accuracy. A typical DSP can execute instructions in less than 100 ns. That capability allows a peak execution rate of 20 MIPS (millions of instructions per second), which is 10 to 20 times the performance of conventional MCUs. DSPs are often rated in million of operations per second (MOPS). The MOPS rating is several times the

MIPS rating. The features of a specific 16-bit DSP, the DSP56L811, include the following:

- 40-MHz, 20-MIPS operation from 2.7V to 3.6V;
- One-cycle multiply-accumulate shifter;
- 16-bit instruction and 16-bit data word;
- Two 36-bit accumulators;
- Three serial ports;
- Sixteen I/Os, two external interrupts;
- Power dissipation of 120 mW at 40 MHz.

The need for real-time processing in several systems is causing control system engineers to evaluate and use DSP technology for the control function. This growing class of functions cannot work effectively with traditional table look-up and interpolate functions to make the control decision. Instead, a multiply-accumulate unit allows state estimator functions to be implemented with an algorithm defining the state. However, as shown in Table 5.2, MCUs have an advantage over DSP units in many areas except real-time operating efficiency. The question that system designers must answer is, "What sample rate and performance level are required for the application?" Also, the possibility of combining the new system with a previously developed or concurrently developed system should be evaluated. Depending on the nature of the system(s) and the availability of control alternatives, a change from MCU to DSP may not be required.

DSP technology is evolving in two ways. First, DSP-like performance can be approached when a time processor unit (TPU) is included in an MCU design, which is possible in higher performance MCUs. The TPU is a programmable microstate machine that addresses requirements for computation and greatly reduces the overhead on the main processor. That allows the main processor to calculate strategy-related items and not have to decide when a specific activity is initiated, such as firing the next sparkplug in an automotive engine control system. In some applications, the main signal processor is a DSP. In other cases, multiple devices can be integrated on the same chip, especially with the capability of integrating an increasing number of transistors per chip. For example, multiple CPUs are used for fault tolerance in safety systems, such as automotive antilock braking systems.

Second, in high-end applications, such as an automotive near-obstacle-detection system or noise-cancellation system, a dedicated 24-bit DSP is used.

Table 5.2
Differences Between DSP and MCU Architectures

Characteristic	DSP	MCU
Ease of programming		Better
Boolean logic		Better
Real-time operating efficiency	Better	
Code maintainability		Better
Peripherals availability		Better

However, code portability is minimal in today's level of DSPs with the strategy closely linked to the I/O and the architecture. Also, generating code for a DSP, in general, is more difficult than generating code for an MCU. However, the multiply-accumulate unit programmed in C requires less effort. As more sophisticated tools are available to deal with DSPs at the behavioral level, the current difficulties in programming fast Fourier transforms (FFTs) and developing filters for DSPs will be simplified.

Figure 5.6 shows a 24-bit DSP and its architecture. The device has program memory, two data memories, and the same kind of instruction set that is common throughout the Motorola MCU family. Using similar instruction sets makes it easier to write code and minimizes the discontinuity for programmers familiar with the MCU family. For example, the multiply, add, move, branch, and bit test are the same. However, the advance features of the chip require new instructions.

A DSP has been used to develop an intelligent sensor for checking the pitch of tapped holes in motor blocks. The sensor replaced a mainframe computer that was checking one hole per minute. By providing DSP power in each sensor, the inspection system increased the rate to 100 holes per minute [8]. The sensor was developed using a DSP design kit that is the size of a credit card. Voltage monitoring, battery backup circuitry, and a watchdog timer are features of the design kit and the DSP-based sensor.

5.4.1 Algorithms Versus Lookup Tables

MCUs use lookup tables to store values that are accessed when the program is running. Algorithms are used to correct for variations from expected results and to implement a control strategy. The speed of accessing the information from a

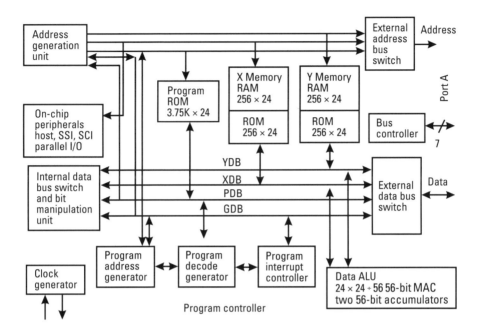

Figure 5.6 24-bit DSP architecture.

table or performing a calculation determines the response time of the sensor input to the MCU/DSP portion of the system. That can be the limiting factor to initiating a change to the output in an MCU/DSP-controlled system.

A control-sensor combination can implement an electronically program-mable trim as an alternative to laser trimming. However, all trimming and cali-bration processes for a sensor require some form of data conversion by the MCU or DSP. The time it takes to perform those conversions by mathematical calculations or get data from a lookup table must be within the control system's ability to respond to the sensed input. In the application, real-time trimming can be implemented to allow adaptive control at the sensor level. That will improve the accuracy of a sensor that has drifted after some time in operation.

5.5   Techniques and Systems Considerations

Increased precision can be obtained for sensors by characterizing devices over temperature and pressure (or acceleration, force, and so forth) and storing cor-rection algorithms in MCU memory. The MCU can convert the measurement to display different units (e.g., psi, kPa, mmHg, or inches of water for pressure measurements). Other techniques use the MCU's capability to improve linearization, to provide PWM outputs for control, and to provide autozero-ing/autoranging. The operating frequency and switching capability of the MCU must be considered in system design. Finally, the MCU's computing capability can be used in place of sensor(s) when sufficient information exists.

5.5.1 Linearization

Sensor nonlinearity can be improved by the use of table lookup algorithms. The variation in sensor signal caused by temperature can also be improved by using an integrated temperature sensor and a lookup table to compensate for temperature effects while linearizing the output, nulling offsets, and setting full-scale gain from information stored in an EEPROM. Lookup tables can be implemented in masked ROM, field programmable EPROM, or onboard EEPROM.

Compensation for nonlinearity and the number of measurements during the test and calibration procedure can be simplified if a nonlinear output corre-lates with a sensor design parameter. For example, a strong correlation was found between the span and the linearity of a pressure sensor with a thin dia-phragm [9]. As Figure 5.7 shows, the nonlinearity increased to almost 5% with the highest span units.

Analytical techniques were investigated to improve the nonlinearity. A polynomial regression analysis was performed on 139 sensor samples ranging from 30- to 70-mV full-scale span to determine the coefficients B_0, B_1, and B_2 in the formula:

$$V_{out} = V_{off} + \left(B_0 + B_1 \cdot P + B_2 \cdot P^2 + B_3 + P^3 ...\right) \qquad (5.1)$$

where B_0, B_1, B_2, and B_3 are sensitivity coefficients. The second-order terms were sufficient for calculations to agree with measured data with a worst case value for calculated regression coefficient = 0.99999. The relationship of those values to the span allowed a piecewise linearization technique with four win-dows to reduce the linearization error of most sensors to less than 0.5%. Those calculations could be included in the MCU lookup table for improved accuracy in an application. Others have also investigated linearization in great detail as a general means to improve sensor accuracy [10, 11].

5.5.2 PWM Control

The PWM output from the MCU can be used to convert an analog sensor out-put to a digital format for signal transmission in remote sensing or noisy

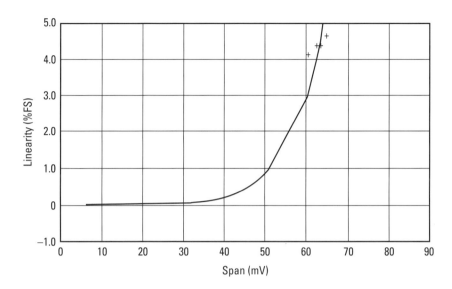

Figure 5.7 Span versus linearity for pressure sensor output.

environments [12]. Figure 5.8 shows the simple, inexpensive circuitry used to create a duty cycle that is linear to the applied pressure. The MCU-generated pulse train is applied to a ramp generator. The frequency and duration of the pulse can be accurately controlled in software. The MCU requires input-capture and output-compare timer channels. The output-capture pin is programmed to output the pulse train that drives the ramp generator, while the input-capture pin detects edge transitions to measure the PWM output pulse width. The pulse width changes from 50 to almost 650 μs for zero to full-scale output for this sensor.

5.5.3 Autozero and Autorange

Combining a sensor and an MCU to perform a measurement that otherwise would be less accurate or more costly than other available alternatives is feasible. The cost of many MCUs is comparable or lower than the micromachined sensors that provide their input signal. For example, a signal conditioned pressure sensor has been combined with a 68HC05 MCU to measure 1.5 inches or less of water with an accuracy of 1% of the full-scale reading [13]. The MCU provides software calibration, software temperature compensation, and dynamic-zero capability. Also, a digital output compatible with the SPI protocol is provided for the pressure measurement.

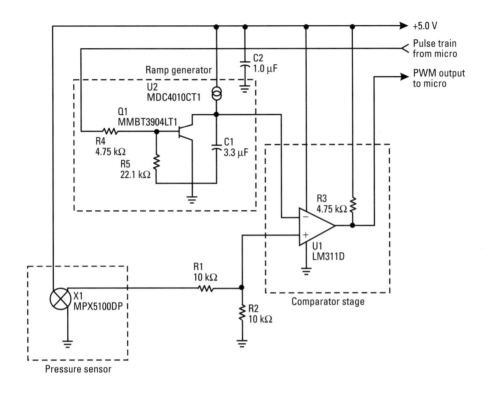

Figure 5.8 PWM output pressure sensor schematic.

Autoreferencing can be performed by the MCU to correct for common-mode errors, especially with low-level signals. Autoreferencing uses the MCU logic and clock signal combined with a DAC and counter. A signal from the MCU initiates the counter. The DAC provides a sample-and-hold and a programmable voltage source. The sensor output is summed with the autoreference correction at the input of an amplifier to obtain a corrected output to the system.

A calibration-free method for pressure sensors has been designed using a calibrated pressure sensor, an MCU with integral ADC, and two additional ICs [14]. As shown in Figure 5.9, two input channels and one output port are used in the calibration portion of the system. The analog switches provide voltages V1 and V2, which are converted and stored in registers in the MCU. The switches are then put in the opposite direction and the new values are stored. The MCU adds the differential results that become an input to the ADC. All errors from the instrument amplifier are canceled in this circuit using the differential conversion. For a measurement that requires only a full-scale accuracy of ±2.5%, the offset of the pressure sensor can be neglected and the system does

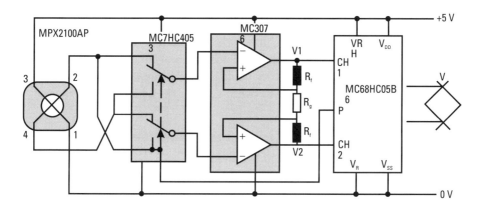

Figure 5.9 Calibration-free pressure sensor system.

not require any calibration procedure. For a ±1% measurement, the full-scale output is set at 25°C.

5.5.4 Diagnostics

One of the more valuable contributions that the MCU can make to the sensor's functionality is the ability to self-test, analyze status, diagnose, and report problems to an operator or other systems that use the sensor's output. The problem could be as simple as a warning for maintenance or calibration or an indication of catastrophic failure that would not allow the system to function when required. For example, an automotive air bag system may operate for years without requiring deployment. However, knowing that the sensor is capable of providing the signal to indicate a crash event is part of the safety the system routinely provides the driver and passengers. In the air bag system, a dash indicator lamp is activated by the MCU to indicate that it has performed a system-ready analysis.

5.5.5 Reducing Electromagnetic Compatibility and Radio Frequency Interference

Radiated RFI and electromagnetic interference (EMI), both the transmitting and receiving of unwanted signals or EMC, of a device are becoming an important design consideration with increasingly higher levels of system integration and higher processing bandwidths. These problems can be reduced at the component level with smaller radiation loops and a smaller number of signal lines. Power management in the MCU can also eliminate some self-induced effects that generate RFI to the sensor element. An example of power management is quiet-time sampling, a circuit technique that performs analog switching when digital switching is not present [6]. Cleaner samples of data are taken by halting high-current, high-frequency activity while the sensor input is being measured. Normal digital processing functions continue once those values are in memory.

5.5.6 Indirect (Computed, Not Sensed) Versus Direct Sensing

Inputs to the MCU can be manipulated to provide additional data for a system. For example, the MCU can use an input pressure signal to provide maximum pressure, minimum pressure, an integrated (averaged) pressure, and time-differentiated pressure data to a system. An accelerometer signal processed by a signal processor can be integrated once to provide velocity and twice to provide displacement information. Also, a single sensor input to an MCU can replace several switches sensing the same parameter and provide programmable switch points for outputs.

The ability to compute rather than sense is among the solutions that MCUs bring to control applications. For example, in three-phase motor control systems, a Hall-effect sensor is used to sense the location of the magnetic field for each phase or the rotor speed in induction motors. An MCU uses the signal to switch output drivers for pulse-width modulated control. The sensors in this system have been a target for cost reduction for many years. Recent solutions have eliminated the sensors by using the MCU to compute the rotor speed and slip angular frequency from other available information, including the primary resistance [15].

5.6 Software, Tools, and Support

Creating a new approach or an alternative to existing control technologies requires much more than the architecture. The software used to program the control portion and the tools that allow the system to be developed are equally important. Portable code (software compatibility) is essential if future end products require migration to a higher performance MCU. Also, keeping the sensor's design and process simple and separate from the MCU allows implementation with other available processors.

One main advantage of using existing MCUs or DSPs is that the development tools already exist and allow the designer to quickly and easily develop both system hardware and software. For example, data entry, program debugging, and programming of an MCU's OTP, EPROM, and EEPROM can be accomplished by utilizing an evaluation module (EVM) (Figure 5.10)

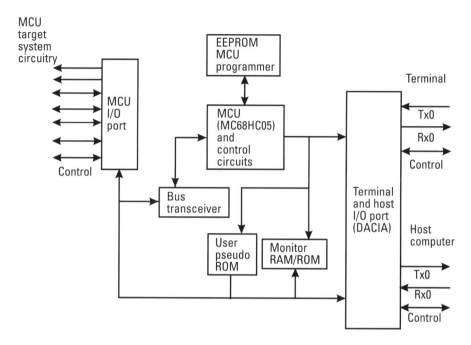

Figure 5.10 MC68HC05EVM development tool for 68HC05 MCU.

interfaced to a host computer such as an IBM-compatible PC or Apple Macintosh® [16]. All the essential MCU timing and I/O circuitry already exist on the EVM to initiate the development process. MCU code can be generated using a resident one-line assembler/disassembler or downloaded to the user program RAM through the host or terminal port connectors. The code can be executed by several debugging commands in the monitor.

Freeware obtained through the World Wide Web can be used in software development. Information, development tools, and documentation established over several years of customer usage are available for many MCUs. As a result, the capability exists to develop the smartest sensor that can be defined today.

5.7 Sensor Integration

The fourth level in advancing integration (see Figure 1.7) is to migrate the developed MOS interface onto the MCU. The basic sensor element is interconnected in close proximity to the enhanced MCU in the same package. Now the sensor itself is the only element that does not need to be MOS process compatible. The fourth level is also the optimum point at which to consider inclusion of any output-drive capabilities required by the specific sensing and control application onto the MCU. This type of power integration could have been accomplished in level III except that standard production MCUs with external drivers could have been more effectively used at the lower levels. At level IV, there is a definite transition from standard-product MCUs into customization for a specific application or market. Therefore, all aspects of the total system should be considered for integration at this level to maximize the benefits of the added design and development investment.

Table 5.3 shows the implications of different process defectivities (defects per unit area) on the net cost of a combined solution at one point in time [7]. Recent industry MOS fabrication processes have typically between 0.5 to 1.0 defects/cm^2 with the best in class at 0.2 defect/cm^2. To demonstrate the costs of combining a sensor with an MCU, consider a small 0.100-inch by 0.100-inch MCU processed at 0.8 defect/cm^2 and a 0.100-inch by 0.100-inch sensor processed at 4.0 defect/cm^2. The example MCU will have a relative cost of 1.05 versus a 0.100-inch by 0.100-inch device processed with zero defects. The sensor will have a relative cost of about 1.29. That gives a combined two-chip relative cost of 2.34. If the sensor and the MCU are combined on a single die using a sensor process that creates 4.0 defects/cm^2, the resulting die would be about 0.128 inch by 0.128 inch and have a relative cost of 2.50. That cost increase of only 1.07 times the combined die cost may be less significant than the added assembly and test costs of two separate devices.

The cost tradeoff changes dramatically as the MCU die size increases. If the same 0.100-inch by 0.100-inch sensor is combined with a larger 0.300-inch by 0.300-inch MCU with relative cost of 15.15, the relative cost of the combined two-chip set rises to 16.44. Combining the two chips into an equivalent die size of 0.311 inch by 0.311 inch and processing at the higher defectivity sensor process will yield a relative cost of 75.28, or 4.58 times the two-chip solution. A more recent (1998) analysis that took into account processing improvements still yielded a 4× cost impact to integrating a bulk micromachined sensor with an MCU. Integrating a (typically) higher defectivity sensor process onto another chip becomes increasingly more costly as the total die size increases. In all cases, adding a sensor to another silicon device will add some cost due to the added silicon area for a fixed batch size (wafer diameter). However, the cost penalty of adding a sensor on a chip decreases if the defectivity of the sensor process decreases. Typically, MCUs use larger wafer size and a higher density process than the sensor, adding to lower yield expectations and higher cost for a combination of the two approaches. As improvements in processing technology occur, the added cost can be traded off against packaging and assembly costs for the two-chip approach.

Table 5.3
Theoretical Relative Die Costs for Sensor Integration

	Best in Class*	Typical MOS Industry*	Expected Level of Sensor Process Development	
Defects/cm^2	0.2	0.8	2.0	4.0
Die size (inches)				
0.100 by 0.100	1.01	1.05	1.14	1.29
0.128 by 0.128	1.70	1.81	2.05	2.50
0.300 by 0.300	10.85	15.15	27.69	65.11
0.311 by 0.311	11.31	16.17	30.89	75.28

*Theoretical 6-inch wafers. Die cost of 1.0 for 0.100 by 0.100 die and zero defects/cm^2.

An interim approach to the fully integrated sensor on an MCU is shown in Figure 5.11. The sensor interface for the accelerometer is initially a standalone circuit that uses the same design rules and process as the MCU. That

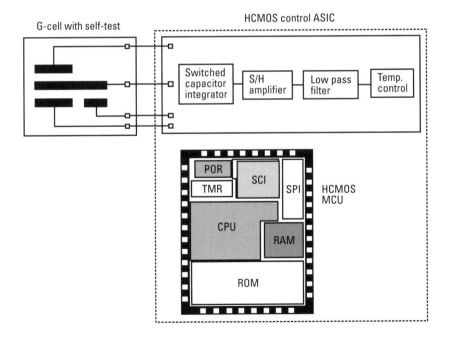

Figure 5.11 Two-chip sensor and HCMOS sensor interface IC.

provides a shorter design cycle and lower initial cost for the sensor. Once the end product (containing the combination of this sensor and the MCU) has achieved market acceptance and a minimum run rate, a phase II design (level III in Figure 1.7) is possible. The sensor interface is treated as a predesigned building block for a fully integrated sensor signal processor using the modular design approach discussed earlier in this chapter.

5.8 Summary

Trends in the development of microelectronics for MCUs, DSPs, ASICs, and FPGAs have been for faster, more complex signal processing and reduced feature size (critical dimensions), into the submicron range. Those goals have led to lower supply voltages (3.3V and less). In addition, customer requirements for increased and more easily programmed memory and reduced development cost and design cycle time are affecting the design and methodology used to design those products. The resulting improvements can be useful in the design of smart sensors and can be cost effective by using other capabilities provided by the MCU, DSP, ASIC, or FPGA. Communicating the data from smart sensors is among the capabilities that can be integrated on chip and is such a critical part that Chapter 7 is dedicated to that aspect of the control logic.

References

[1]   Jacobsen, E., "Signal Conditioning a Pressure Sensor With a Field-Programmable Analog Array," *Sensors*, Nov. 1997, pp. 81–86.

[2]   Najafi, N., and K. Wise, "An Organization for Interface for Sensor-Driven Semiconductor Process Control Systems," *IEEE Trans. Semiconductor Manufacturing*, Vol. 3, No. 4, Nov. 1990, pp. 230–238.

[3]   Sibigtroth, J. M., *Understanding Small Microcontrollers*, Motorola Technical Bulletin M68HC05TB/D Rev. 1, 1992.

[4]   "FRAM Makes a Comeback," *Portable Design*, Oct. 1996, pp. 14–16.

[5]   *M68HC05 Microcontroller Applications Guide*, Motorola M68HC05AG/AD.

[6]   Benson, M., et al., "Advanced Semiconductor Technologies for Integrated Smart Sensors," *Proc. Sensors Expo '93*, Oct. 26–28, 1993, pp. 133–143.

[7]   Frank, R., J. Jandu, and M. Shaw, "An Update on Advanced Semiconductor Technologies for Integrated Smart Sensors," *Proc. Sensors Expo West*, Anaheim, CA, Feb. 8–10, 1994, pp. 249–259.

[8] "MiniKit 56001 Provides DSP Power to Low-Volume Products," *DSP News,* Vol. 6, Issue 3, 3Q93, Motorola, 1993, p. 5.

[9] Derrington, C., "Compensating for Nonlinearity in the MPX10 Series Pressure Transducer," AN935 in *Sensor Device Data/Handbook,* DL200/D Rev. 4, 1998.

[10] Hille, P., R. Hohler, and H. Strack, "A Linearisation and Compensation Method for Integrated Sensors," *Sensors and Actuators* A 44, 1994, pp. 95–102.

[11] Heintz, F., and E. Zabler, "Application Possibilities and Future Chances of 'Smart' Sensors in the Motor Vehicle," 890304 in SAE *Sensors and Actuators* SP-771, 1989.

[12] Jacobsen, E., and J. Baum, "Using a Pulse Width Modulated Output With Semiconductor Pressure Sensors," AN1518 in *Sensor Device Data/Handbook,* DL200/D Rev. 4, 1998.

[13] Ajluni, C., "Pressure Sensors Strive to Stay on Top," *Electronic Design,* Oct. 3, 1994, pp. 67–74.

[14] Burri, M., "Calibration Free Pressure Sensor System," AN1097 in *Sensor Device Data/Handbook,* DL200/D Rev. 4, 1998.

[15] Kanmachi, T., and I. Takahashi, "Sensor-Less Speed Control of an Induction Motor," *IEEE Industry Applications,* Jan./Feb. 1995, pp. 22–27.

[16] Frank, R., "Two-Chip Approach to Smart Sensors," *Proc. Sensors Expo 1990,* pp. 104C-1–104C-8.

6

Communications for Smart Sensors

Behold the people is one, and they all have one language; and this they begin to do: and now nothing will be restrained from them, which they have imagined to do.

—Genesis 11:6

6.1 Introduction

The increasing interest in smart sensors is a direct result of the need to communicate sensor information in distributed control systems. Unfortunately, numerous protocols have been defined for data communication in control systems. Industry standards have been and are being developed for various applications. Within a given market segment, several proposals are vying for acceptance. A few of these protocols have already been implemented in silicon hardware. This chapter provides background information, identifies many of the proposed protocols, and describes silicon chips that have been developed to support system designs. (Chapter 12 goes into the details of the newest IEEE1451 standards developed specifically for sensors.)

6.2 Definitions and Background

6.2.1 Definitions

Data communication from sensor inputs must be analyzed, with the recognition that the sensor is a part of an entire control system. Several industry committees and individual companies have expended considerable effort to generate acceptable protocols that will support the requirements for distributed control. System functionality and the ability to use available silicon hardware, software, and development tools must be considered in the selection of a protocol for a given application.

Several terms and definitions are used to describe the communication of system information. A *protocol* is an agreed-on set of rules for communications. The *bus* connects internal or external circuit components and can be serial or parallel; the serial approach is more common. A *device-level bus* connects basic control elements, the sensors and actuators, to a host controller. *Multiplexing* (MUX) is the combining of several messages for transmission over the same signal path. Access to the bus is obtained through an *arbitration* process. Bit-by-bit arbitration is also called *collision detection*. *Contention* is the ability to gain access to the bus on a predetermined priority. *Latency* is guaranteed access (with maximum priority) within a defined time. A *deterministic* system can predict the future behavior of a signal. The *data link controller* (DLC) is the silicon implementation of the protocol that handles all communications requirements.

The system goal of interoperability permits sensors or actuators from one supplier to be substituted for one from another manufacturer. Different types of networks are shown in Figure 6.1. Star, ring, and linear (tree, multidrop) are common *topologies*, that is, methods of connecting data lines to system nodes. The use of master-slave relationships, although an important part of control networks, is giving way to the use of more distributed intelligence in many applications.

The 4- to 20-mA standard of the Instrument Society of America (ISA) has been the analog data transmission standard for over 30 years. However, digital techniques can provide more system functionality and increased noise immunity for signal transmission. Many digital interface formats already exist for point-to-point or multidrop communications, as shown in Table 6.1 [1]. These protocols have seen limited use in computer-controlled systems. The EIA-485 is most often used for digital field bus applications. It is the physical layer in Profibus, Topaz, and Bitbus protocols. The EIA-485 is a balanced-line (differential) data transmission over a single twisted-wire pair [1]. However, examining the different market segments, an even wider variety of standards exists.

6.2.2 Background

The International Organization for Standardization (ISO) has defined the Open Systems Interconnection (OSI) model that describes seven layers for data networking. The layers and their functions are shown in Figure 6.2 [2].

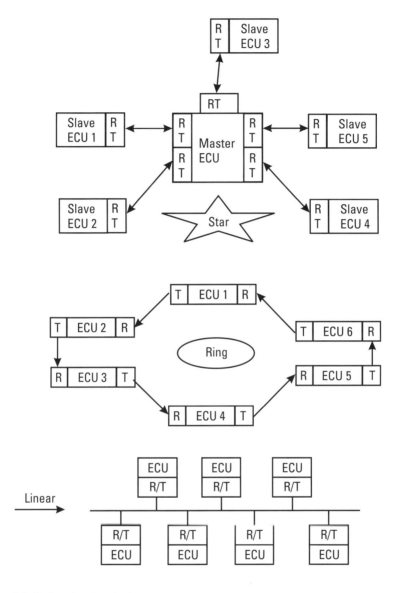

Figure 6.1 Various bus topologies.

Available and proposed standards typically use all or a portion of the layers to define the standard.

In a distributed system, a node contains the sensor (or actuator), the hardware for local computations, and the network interface [3]. The term *sensor bus*

Table 6.1
Point-to-Point and Multidrop Standard

Parameter	EIA-232-C	EIA-432-A	EIA-422-A	EIA-485
Mode of operation	Single ended	Single ended	Differential	Differential
Number of drivers and receivers allowed	1 driver, 1 receiver	1 driver, 10 receivers	1 driver, 10 receivers	32 drivers, 32 receivers
Maximum cable length (ft)	50	4,000	4,000	4,000
Maximum data rate (bps)	20k	100k	10M	10M
Maximum common-mode voltage	± 25V	±6V	6V, −0.25V	12V, −7V
Driver output	±5V min.; ±15V max.	±3.6V min.; ±6V max.	± 2V min.	±1.5V min.
Driver load	±3 kΩ–7 kΩ	450Ω min.	100Ω min.	60Ω min.
Driver slew rate	30V/μs max.	Externally controlled	NA	NA
Driver output short circuit current limit	500 mA to Vcc or GND	150 mA to GND	150 mA to GND	150 mA to GND 250 mA to −8V or 12V
Receiver output resistance (on)	NA	NA	NA	120 kΩ
high Z state) (off)	300Ω	60Ω	60Ω	120Ω
Receiver input resistance	3–7 kΩ	4 kΩ	4 kΩ	12 kΩ
Receiver sensitivity	± 3V	± 200 mV	± 200 mV	± 200 mV

is frequently used when sensors are connected in a system using a multiplexed bus offering data link–level multiplexing and allowing packets from different senders to be sent to different receivers. A higher level sensor network uses all layers of the OSI model to provide more information and simplify the user's system design and maintenance activity.

6.3 Sources (Organizations) and Standards

Several universities have developed standards for communication. For example, the University of Michigan has proposed the Michigan parallel standard (MPS)

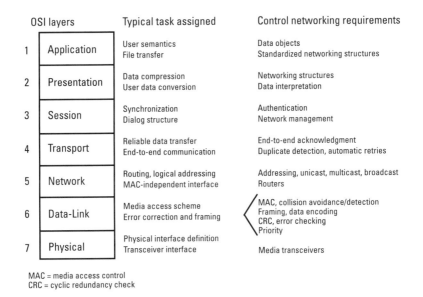

Figure 6.2 ISO's Open Systems Interconnection (OSI) model.

[4]. Researchers at Delft University of Technology have developed a serial communications protocol [5]. The Integrated Smart Sensor (IS^2) bus is a mixed analog/digital two-line serial bus interface. However, a protocol requires more than definition and demonstration hardware. Accepted usage by a number of manufacturers ultimately determines the real standards that may, in fact, be de facto standards. University-developed protocols may achieve acceptance with the adaptation of industrial users, but the process takes longer than utilizing existing industry-sponsored standards. The focus of this chapter is standards that industry already supports. Table 6.2 lists the more common standards, including those developed by universities, manufacturers, and standards organizations.

6.4 Automotive Protocols

Due to its large volumes, the automotive segment has driven specifications, resulting in the actual implementation of protocols in several products. In automobiles, information from one sensor and/or data from one system can be communicated with other systems using multiplex wiring to reduce the number of sensors and the amount of wire used in a vehicle. Two predominant protocols have emerged as standards, but several other protocols exist for

Table 6.2
Protocols and Sponsors in Various Market Segments

Protocol Automotive	Sponsor
J-1850	Society of Automotive Engineers (SAE)
J-1939 (CAN)	SAE
J1567 C^2D	SAE (Chrysler)
J2058 CSC	(Chrysler)
J2106 Token Slot	SAE (GM)
CAN	Robert Bosch GmbH
VAN	ISO
A-Bus	Volkswagen AG
D^2B	Philips
MI-Bus	Motorola
TTP	University of Wien, Austria
Industrial	
HART	Rosemount and HART Communication Foundation
DeviceNET™	Allen-Bradley
Smart Distributed Systems (SDS™)	Honeywell
SP50 fieldbus	ISP + World FIP = Fieldbus Foundation
SP50	IEC/ISA
LonTalk™	Echelon Corp.
Profibus	German DIN
ASI bus	ASI Association
InterBus-S	Phoenix Contact and InterBus-S Club
Seriplex	Automated Process Control and Seriplex Technology Organization
SERCOS	SERCOS N.A.
IPCA	—
ARCNet	Datapoint Corp. and ARCNet Trade Association
Building/office automation	
BACnet	Building Automation Industry
LonTalk™	Echelon Corp.
IBIbus	Intelligent Building Institute
Batibus	Merlin Gerin (France)

Protocol		
Building/office automation		
Elbus	Germany	
CAB	Canada	
Home automation		
X-10	X-10 Corp.	
Smart House	Smart House L.P.	
CEBus	EIA	
LonTalk™	Echelon Corp.	
Michigan Parallel Standard	University of Michigan	
Integrated Smart-Sensor Bus	Delft University of Technology	
Time-triggered protocol	University of Wien, Austria (Automotive)	

DeviceNET™ is a trademark of Allen-Bradley Company, Inc. SDS™ is a trademark of Honeywell, Inc. LonTalk™ is a trademark of Echelon Corporation.

specific manufacturers' applications, as shown in Table 6.2. The Society of Automotive Engineers (SAE) has established SAE J1850 as the standard for multiplexing and data communications in U.S. automobiles. However, data communications for trucks and On-Board Diagnostics II (OBDII) are based on the Controller Area Network (CAN) protocol developed by Robert Bosch GmbH.

The SAE Vehicle Network for Multiplexing and Data Communications (Multiplex) Committee has defined three classes of vehicle networks: class A, class B, and class C [6]. Class A is for low-speed applications such as body lighting. Class B is for data transfer between nodes to eliminate redundant sensors and other system elements. Class C is for high-speed communications and data rates typically associated with real-time control systems. Table 6.3 summarizes the data rates and latencies for the three classes.

6.4.1 SAE J1850

SAE J1850 was approved as the standard protocol for U.S. automakers in 1994. J1850 defines the application, data link, and physical layers of the OSI model. J1850 specifies two implementations: a PWM version and a variable pulse width (VPW) version. The differences are indicated in Table 6.4. SAE

Table 6.3
Automotive Network Classes

Class	Type	Data Rate	Latency
A	Low	1 Kbps to 10 Kbps	20–50 ms
B	Medium	10 Kbps to 100 Kbps	5–50 ms
C	High	10 Kbps to 1 Mbps	1–5 ms

J1850 is supported by a number of other SAE specifications, which are referenced in J1850, such as the message strategy document, J2178.

In addition to the SAE bit-encoding techniques specified in Table 6.4, other common techniques include 10-bit nonreturn to zero (NRZ), Bit-stuf NRZ, L-Man (Manchester), E-Man, and modified frequency modulation (MFM). These techniques differ based on variable synchronizing, arbitration, transitions per bit, maximum data rate, oscillator tolerance, and integrity. SAE J1850 network access is nondestructive prioritized by bit-by-bit arbitration for either protocol option [6]. A frame is defined as one complete transmission of information. Within the frame, the header contains information regarding the message priority, message source, target address, message type, and in-frame response. For SAE J1850, each frame contains only one message, and the maximum length for a frame is 101 bit times. A power reduction or sleep mode occurs at a node if the bus is idle for more than 500 ms. Wakeup occurs with any activity on the bus.

6.4.2 CAN Protocol

CAN, a serial communications protocol developed by Robert Bosch GmbH, was originally designed for automotive multiplex wiring systems, especially

Table 6.4
Protocol Options in SAE J1850

Feature	1- and 3-Byte Headers	1- and 3-Byte Headers
Bit encoding	PWM	VPW
Bus medium	Dual wire	Single wire
Data rate	41.7 Kbps	10.4 Kbps
Data integrity	CRC	CRC

high-speed data communications. CAN supports distributed real-time control with a high level of security and message integrity. It has also become attractive for use in lower speed and other distributed control applications.

The original CAN specification was announced in the 1980s. A revision, CAN 2.0, was announced in 1991. CAN 2.0 consists of an A part and a B part. Part A is known as CAN 2.0 A, CAN 1.2, and BasicCAN. Part A specifies an 11-bit identifier field, includes no specification for message filtering, and has a layered architecture description based on Bosch's internal model. Part B enhancements include an extended 29-bit identifier field, some message filtering requirements, and layer description based on ISO/OSI reference model. The 29-bit identifier field allows the automotive standard protocol, SAE J1850, 3-byte headers to be mapped into the CAN identifier field. However, minimum CAN compliance is established by conformance to CAN 2.0 A only [7].

CAN is a multimaster protocol that allows any network node to communicate with any other node on the same network. Any node can initiate a transmission once it has determined that the network is idle. CAN properties include user-defined message prioritization. CAN is actually a nondeterministic system, but a guaranteed maximum latency for highest prioritization can be calculated. Lower priorities are determined on a statistical basis. CAN utilizes carrier sense, multiple access/collision resolution (CSMA/CR) for nondestructive collision resolution. The arbitration technique is bitwise and results in the highest priority message being transmitted with low latency time. The flexible system configuration allows user options that have been led to CAN-developed systems in automotive and industrial applications [7].

The message format for CAN is a fixed-format frame with a variable number of data bytes. Zero to 8 data bytes are permissible. The minimum data frame length is 44 bit times. Four message types are defined: data frame, remote frame, error frame, and overload frame. The messages are routed to receiving nodes through the use of message identifiers and message filtering. Functional addressing allows multiple nodes to act on a single message. However, nothing in the hardware prevents the user from using physical addressing to achieve node-to-node addressing. Part of the error detection scheme is the acceptance of every message by all nodes or no nodes. Even if a component on the network is not concerned with the message that is transmitted, it must still receive the message, check the cyclic redundancy check (CRC), and acknowledge (ACK) acceptance. The data frame, shown in Figure 6.3, is the most widely used frame. Priority is established in the identifier field based on the user's messaging strategy [7].

Data is requested through the use of a remote frame and transmitted through the data frame. The remote frame contains no data field and the

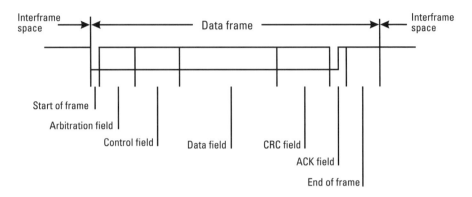

Figure 6.3 CAN data frame.

remote transmission request (RTR) bit is sent recessive. Otherwise, the remote frame is identical to the data frame [8].

Error detection and error signaling are possible because CAN has a variety of built-in error counters that help contain errors and prevent nodes that have failures from restricting communications on the bus. A faulty transmitting node always increases its error counter more than other nodes. Therefore, the faulty node becomes the first to go "bus off." Automatic retransmission of corrupted messages minimizes the possibility of an undetected message. CAN also has the ability to distinguish between temporary errors and permanent node failures, which makes it ideal for the high-noise environments that occur in automotive and industrial applications [7]. The error message frame is shown in Figure 6.4. Bit error, stuff error, CRC error, form error, and acknowledge error can be detected. The error frame contains the superposition of error flags sent by various nodes on the network and the error delimiter. Both passive and active error codes are possible in CAN, depending on the status of the node [8].

The CAN protocol does not address the physical layer requirements. The transceiver is not specified, but rather it is left to the user's network

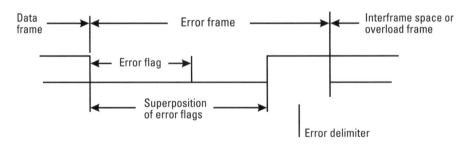

Figure 6.4 CAN error frame.

characteristics to define the requirements. (Bit timing rules in the CAN specification must be met.) Accepted physical media include, but are not limited to, twisted pair (shielded or unshielded), single wire, fiber-optic cable, and transformer coupled to power lines. RF transmitters are also being developed by some users for CAN systems. Acceptable transmission rates range from 5 Kbps to 1 Mbps. The most widely used implementation to date is a twisted-pair bus with NRZ bit formatting. Twisted-pair CAN drivers that simplify system design are available [7].

6.4.3   Other Automotive Protocols

As shown in Table 6.1, other automotive protocols have been developed, some of which are covered by SAE specifications. The implementation of those alternatives requires a volume user that obtains benefit by using the alternative protocol, as well as a semiconductor manufacturer willing to implement the protocol in silicon hardware. For comparison, the Automotive Bit-Serial Universal Interface System (A-bus) and the vehicle area network (VAN) are discussed.

A-bus was developed by Volkswagen AG [6]. Error detection is by bit only. Both single wire and fiber optics can be used for the transmission media. The maximum bus length is not specified but is typically 30m. The maximum data rate of 500 Kbps is considerably higher than the rate specified in SAE J1850.

VAN, developed by French automaker Renault, is being considered as an ISO standard. The transmission medium is twisted pair. The maximum data rate is user definable. A maximum of 16 nodes is possible with a maximum bus length of 20m. Bit encoding is one of two variations on Manchester coding. The Manchester coding technique encodes a 1 with a high level for the first half of the bit time, and a 0 is encoded with a low level for the second half of the bit time. Consequently, a maximum of two transitions per bit time is possible.

The Technical University of Wien in Austria has developed a time-triggered protocol (TTP). The protocol provides information about the future (i.e., a priori knowledge) behavior of the system through timing algorithms. All system activities are triggered by the progression of real time. The architecture requires clock synchronization by a global time base. TTP supports high-speed distributed control systems, where fault-tolerant operation is critical.

A consortium of German and French automakers have developed an operating system (OS) standard for automotive electronics. The OSEK (German acronym for Open System for Automotive Electronics) OS is an extension of the front end of the MCU hardware [9]. It is intended to provide a structured design with additional higher software levels encapsulating lower levels to reduce the software complexity for distributed control systems in vehicles. The standard allows standardized software components to be used across multiple platforms to reduce development cost and time to market.

OSEK was developed to be independent of the communication protocol, so it is compatible with the CAN, J1850, and other automotive protocols shown in Table 6.2. The combination of application-specific OS, protocol, and system hardware, including smart sensors, will be required to improve the efficiency, meet emissions and safety standards, and improve the overall driving experience in future vehicles.

6.5   Industrial Networks

The power of a distributed multiplex system is easily demonstrated in factory automation. In many cases, users have completed wiring a multiplex system installation with one or two people in a single day, a task that previously took a crew of electrical technicians several days to wire. Also, these installations work successfully the first time, resulting in a considerable cost reduction. Wiring for a multiplex system consists of a twisted pair of wires, power, and ground, which greatly simplifies the interconnection process. Once a system strategy is developed, nodes can be added or moved easily without reengineering the system. The nodes can include sensors, as well as valves, motors, and lighting loads. The key is an open standard and plug-and-play capability [7].

The industrial market has even more proposed and developing standards than the automotive market. *Fieldbus* is the term for a nonproprietary digital two-way communications standard in the process automation industry. The fieldbus specification will define the application, data link, and physical layers of the ISO model with some layer-4 services defined. Figure 6.5 illustrates a typical industrial application, with fieldbus as the highest performance level and a sensor bus at the lowest level [10]. Fieldbus has not been completed, and semiconductor products are not available to implement the control nodes. Two protocols that are attracting a number of industrial users based on available silicon products are CAN and LonTalk™.

6.5.1   Industrial Usage of CAN

CAN has been adopted for use in industrial applications by manufacturers such as Allen-Bradley and Honeywell in the DeviceNET™ system and SDS™, respectively. Those communication networks were developed as simpler, lower cost alternatives to fieldbus, which is being developed as an industrial standard.

Typical configuration of large open control system

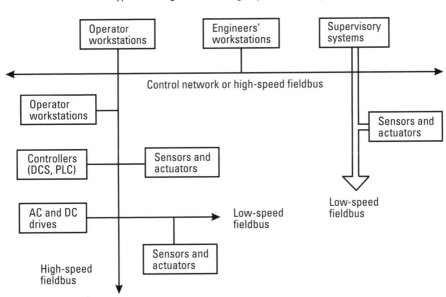

Figure 6.5 Fieldbus control system architecture. (*After:* [10].)

However, fieldbus is intended to handle a larger amount of data. All three networks are designed for real-time operation, but each handles a different amount of transmitted data.

The CAN protocol is used to implement both SDS™ and Device-NET™, but the two systems are not interoperable. The difference between them is in the physical layer, which is not defined in the CAN specification, allowing users to implement different approaches.

6.5.2 LonTalk™ Protocol

The LonTalk™ protocol, developed by Echelon Corporation, has received considerable support for several industrial and consumer applications. The protocol defines all seven layers of the OSI model. It uses differential Manchester coding with data packet length of 256 bytes and can operate up to a maximum speed of 1.25 Mbps. The arbitration is predictive carrier sense, multiple access (CSMA) and has collision detection and priority options. The LonTalk™ protocol can be used to support sensor network to fieldbus requirements [11]. A LonWorks™ system works on a peer-to-peer basis and does not require a host central processing unit (CPU).[1] Events occurring at a particular device are communicated directly between modules. LonWorks control network technology has been selected as a standard by Semiconductor Equipment Materials International (SEMI). SEMI standard E-61 specifies LonWorks as a sensor bus for connecting simple as well as complex sensors and other equipment in semiconductor manufacturing.

6.5.3 Other Industrial Protocols

Other industrial protocols listed in Table 6.2 offer advantages to distributed control systems. A brief summary of the most widely known, frequently mentioned, and better-documented protocols will conclude this section. These include the highway addressable remote transducer (HART), fieldbus, process fieldbus (Profibus), SERCOS, Topaz, and ARCNet™ protocols.

Rosemount developed the HART protocol in the late 1980s. It has a dedicated users group (HART Communication Foundation) with several participating companies. The open multiple master protocol is compatible with the 4- to 20-mA current loop and also allows digital communication [12].

The recent combination of the Interoperable Systems Project Foundation (ISPF) and WorldFIP North America into the Fieldbus Foundation has the potential to complete a fieldbus specification that has been in development for over 10 years. Fieldbus defines the user layer in the OSI model, making it more extensive than protocols that address only the physical, data link, and/or application layers (e.g., CAN) [13].

Profibus is a German Deutsches Industrial Norm (DIN) standard that supports a data rate up to 1.5 Mbps. The deterministic system uses a master-slave hierarchy and has a capacity of 127 nodes. Arbitration is handled by command response. EIA-485 is used as the physical layer [13]. A serial, real-time communication system (SERCOS) network typically consists of eight nodes per ring. It operates at a speed of up to 1 Mbps and for a distance of up to 40m at that speed. It is a deterministic multimaster network with command response and broadcast for arbitration. The physical layer is fiber-optic [12].

Topaz can handle up to 255 nodes and operate at a speed of 1 MHz. At 1 km, the speed decreases to 500 Kbps. The multimaster system is deterministic and uses token passing for arbitration. The physical layer uses EIA-485 interface [12].

ARCNet™ (Attached Resource Computer Network), a trademark of Datapoint Corp., is a deterministic token-passing protocol [14]. In an

---

1. LonWorks™ is a trademark of Echelon Corporation.

ARCNet™ system, all nodes have equal access to the network, eliminating transmission collisions on busy networks. The ARCNet™ protocol automatically reconfigures the network without software intervention if a node is added or deleted.

Other specialty buses, including ControlNet, Genius I/O, and Sensoplex, have been developed by industrial control manufacturers. Table 6.5 lists the attributes of 14 industrial bus systems [12]. Based on the broad range of capabilities, it is easy to see why so many different bus types have been developed.

6.6 Office/Building Automation

The BACnet protocol has been developed by the building automation industry. Ethernet, ARCNet™, MS/TP, and LonWorks™ are among the networks that could communicate on the proposed BACnet-compatible system being developed by the American Society of Heating, Refrigeration, and Air-Conditioning Engineers (ASHRAE). Building energy management systems standards are also being developed. In addition, the Automatic Meter Reading Association (AMRA) is developing a standard for automatic meter reading. IBIbus has been developed by the Intelligent Building Institute [15].

Smart office buildings offer a high degree of automation. Figure 6.6 shows the interconnection of several systems [15]. In offices, nodes sense changes in the environment and send status and control messages to other nodes in response to those changes. Power nodes open or close dampers, change fan speed, and make other adjustments based on that information.

Table 6.5
Comparison of Industrial Communications Networks

Attribute	Sensor Bus	Device Network	Fieldbus
Network name	SDS™	DeviceNET™	WorldFIP, ISP
Target devices	Sensors	Pushbuttons, sensors, switches, drives, starters, valves	Smart transmitters, flowmeters, servo valves
Data packet	Typically 1 byte	Up to 8 bytes	Up to 1,000 bytes
Data rate	≤CAN	Up to 500 Kbps	Up to 1–2.5 Mbps
Data quantity	Limited data	Moderate data	Large amounts of data
Interoperability	Vendor specific	Open	Open

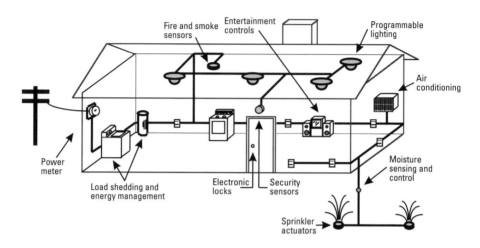

Figure 6.6 Building automation.

Other aspects of these systems are self-diagnostics, data logging, fire detection and sprinkler systems, energy usage monitoring, and security systems.

6.7 Home Automation

The computer control of future homes is the goal of the Smart House project. Among the likely candidates for interfacing to the network are the heating, ventilation, and air conditioning (HVAC) system; water heater; range; security; and lighting [16]. Remote meter reading and demand-side management by utility companies are among the driving forces for home applications. Protocol acceptance in this consumer environment is contingent on achieving low cost/node and ease of operation. The speed of those systems will range from low to high, depending on the devices connected to the system. Both the message size and the message protocol complexity will be medium.

Over two decades ago, X-10 Corporation developed the X-10 protocol for homes, which has been used extensively for lamp and appliance controls [17]. More recently, the Smart House Applications Language (SHAL) was developed and includes over 100 message types for specific functions. Dedicated multiconductor wiring is required in the home. The system can address 900 nodes and operates at a maximum of 9.6 Kbps. Two additional contenders in this arena are CEBus and LonTalk™.

6.7.1 CEBus

The consumer electronics bus (CEBus) was initiated by the Consumer Electronics Group of the Electronic Industries Association (EIA). CEBus provides both data and control channels and handles a maximum of 10 Kbps. It has a growing acceptance in the utility industry [13].

6.7.2 LonTalk™

The acceptance of LonTalk™ in building automation, the availability of full OSI layers, and interoperability are among the reasons that it is well suited for the home automation environment. Simplicity and ease of installation are also important attributes, especially for add-on equipment. Widespread acceptance in other markets will also drive the learning curve for cost reduction that this market requires.

6.8 Protocols in Silicon

A few of the protocols discussed in this chapter have been implemented in silicon hardware and are available from multiple sources. In some cases, the protocol is a standalone integrated circuit. However, a protocol integrated into an MCU provides computing capability and a variety of system features that affect the cost effectiveness of individual system nodes.

6.8.1 MCU With Integrated SAE J1850

A standard protocol provides sufficient volume potential for semiconductor manufacturers to design a variety of integrated circuits. In many cases, the highly integrated solution provides the lowest cost per node. For example, Motorola's MC68HC705V8, shown in Figure 6.7, integrates the J1850 communication protocol, on-chip voltage regulation, the physical layer interface, the J1850 digital aspect (VPW version), and other systems-related functions. The integrated solution has many features that make system design easier and reduce both component count and system cost.

The heart of the J1850 chip is Motorola's data link controller (MDLC), illustrated in Figure 6.8 [18]. The MDLC handles all communication duties, including complete message buffering, bus access, arbitration, and error detection [19]. Other J1850 versions have been implemented by a number of suppliers [20].

Figure 6.7 Systems integration in MC68HC705V8 IC. (Courtesy of Motorola, Inc.)

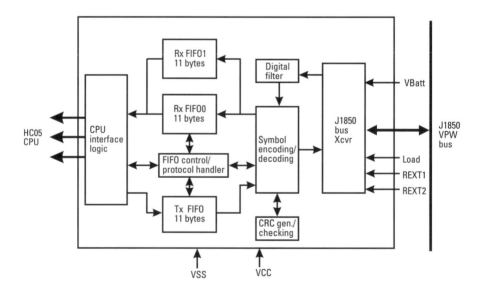

Figure 6.8 SAE J1850 data link controller.

6.8.2 MCU With Integrated CAN

Several semiconductor manufacturers have implemented the CAN protocol on a variety of microcontrollers. The integrated CAN solution contains a variety of I/O, memory, and other system features. One key point of the CAN protocol is that, although a set of basic features must be implemented in any CAN device, a number of extended features may also be implemented in silicon, depending on the intended target application. The implications must be understood when interoperability and interchangeability are being considered. A number of semiconductor manufacturers have designed and offer several solutions. In fact, the broad product availability is among the reasons that industrial users are choosing CAN [7].

An example of the CAN protocol implemented in an MCU is provided by Motorola's M68HC05 microcontroller family. This CAN implementation conforms to the CAN 2.0 part A specification. All Motorola CAN (MCAN) devices include automatic bus arbitration, collision resolution, transmission retry, digital noise filtering, message ID filtering, frame generation and checking, CRC generation and checking, as well as single transmission and multiple receive frame buffers [20]. As shown in Figure 6.9, the MCAN module contains full message transmit and receive buffers and limited message filtering capability [21]. The integrated bus interface includes input comparators and CMOS output drivers. However, an external transceiver may be necessary, depending on the user's application.

The CAN module is totally CPU independent and can be integrated with a CPU core, other MCU hardware, and I/O functions into a single chip. Figure 6.10 shows the block diagram of a 32-bit reduced instruction set computer (RISC) microcontroller with two integrated CAN 2.0 B (Toucan) modules [22]. The chip has two third-generation timer processing units that can handle fuel and spark computations without interrupting the main processor. Each unit has its own 32-bit MicroRISC engine, capable of processing 20 million instructions per second, and features its own onboard scheduler and 16 timer channels. Additional peripherals include two queued ADC modules, a queued serial multichannel module with two high-speed universal asynchronous receiver transmitters and a queued serial port, an 18-channel modular input/output subsystem (MIOS1), and a system-bus interface control block with general-purpose I/O support. This complex chip was designed specifically for automotive powertrain control.

For automakers to achieve complete system functionality, a range of products must be available. Table 6.6 lists other lower cost 8-bit MCUs with a variety of features [23]. Example software drivers available for these units

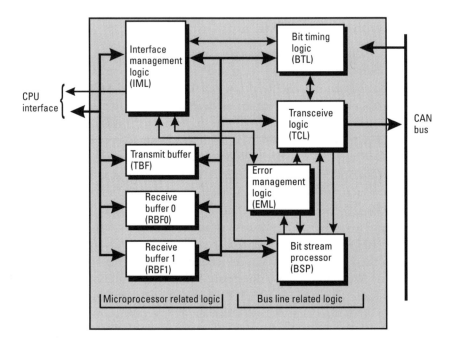

Figure 6.9 CAN module.

provide an interface between application software residing in the MCU ROM (or EPROM) and the CAN module. The routines initialize the CAN module, transmit messages previously stored in RAM, and automatically handle received messages [7].

Memory, in RAM, ROM, EPROM, or EEPROM, is required for proper CAN operation. The smaller X4 version includes 4,096 bytes of EPROM (one-time programmable version) or ROM and 176 bytes of RAM. The X16 and X32 versions have 15,120 bytes of (EP)ROM, 256 bytes of EEPROM, and 352 bytes of data RAM for more complex programs. The EEPROM on the X16 also provides flexibility for addressing and programming over the network. A user can manufacture a number of identical modules and put the ID in EEPROM when each is installed, to minimize inventory costs. The A/D allows easy interface for sensors, and the PWM module can be used for digital speed control of power devices. The integrated timers can be used as part of the program control. Also, the buffering of the CAN module requires CPU intervention for each message received or transmitted. Due to the additional functionality designed into the MC68HC05X16, the CAN module occupies less than 20% of the active area of the chip, a feature that makes the integrated function very cost effective [24].

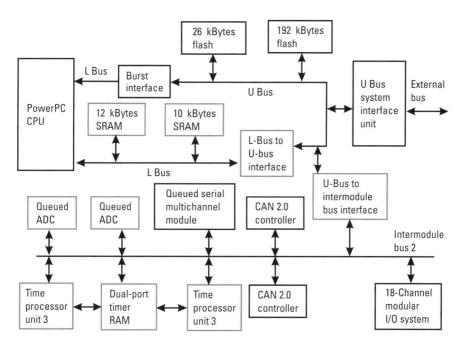

Figure 6.10 Block diagram of an MPC555 PowerPC™ MCU, a chip that integrates two CAN modules and a number of additional system features. (PowerPC™ is a trademark of International Business Machines.)

6.8.3 Neuron® Chips and LonTalk™ Protocol

The LonTalk™ protocol is an integral part of a Neuron IC, such as Motorola's MC143150 or MC143120.[2] The Neuron IC is a communications and control processor with an embedded LonTalk™ protocol for multimedia networking environments, in which received network inputs are used to control processor outputs. As shown in Figure 6.11, the MC143150 has three CPUs: one for the media access control (MAC) processor, one for the network processor, and one for the application processor [7].

The MAC processor handles layers 1 and 2 of the ISO protocol stack. That includes driving the communications subsystem hardware as well as executing the collision avoidance algorithm. This processor communicates with the network processor using network buffers located in shared memory [12].

The network processor implements layers 3 through 6 of the ISO stack. Network variable processing, addressing, transaction processing, auth-

2. Neuron® is a registered trademark of Echelon Corporation.

Table 6.6
MCUs With Integrated CAN Modules and Additional Features

Feature	X4	X16	X32
MC68HC05 CPU	Yes	Yes	Yes
Basic CAN protocol (Rev. 2.0 part A)	Yes	Yes	Yes
8-bit ID mask register for optional message filtering	Yes	Yes	Yes
1 Tx message buffer	Yes	Yes	Yes
2 Rx message buffers	Yes	Yes	Yes
Bytes of user program (EP)ROM	4,096	15,120	31,248
Bytes of EEPROM	—	256	256
Bytes of use data RAM	176	352	352
Programmable input/output pins, ( ) input only	16	24 (8)	24(8)
Port B wired OR interrupt operation	—	Yes	Yes
Serial communication interface (SCI)	—	Yes	Yes
8-channel ADC	—	Yes	Yes
16-bit timer with _ input capture + _ output compares	1 + 1	2 + 2	2 + 2
2 pulse length modulation systems	—	Yes	Yes
Chip operating properly (COP) watchdog (with COP enabled on reset option)	Yes	Yes	Yes
EPROM and EEPROM protection	—	Yes	Yes
Package (number of pins)	28	64/68	64/68

entication, background diagnostics, software timers, network management, and routing functions are handled by this unit. In addition to communication with the MAC processor, it communicates with the application processor through application buffers.

The application processor executes the code written by the user and the operating services requested by user code. The primary programming language is Neuron C, a derivative of the C language. Enhancements in Neuron C for distributed control applications are supported in firmware. As a result, the user can write complex applications with less program memory.

Both Neuron® chips include a direct-mode transceiver and 25 application I/O models. Other integrated functions include the network communication port; several I/O functions, including two timers/counters, clocking and

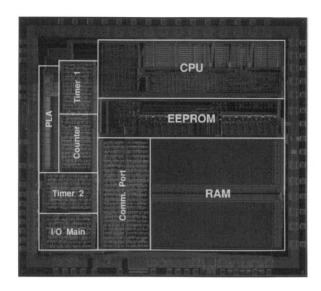

Figure 6.11 Block diagram of MC143150 Neuron® chip.

control capability; and (for the MC143150) 2 KB of RAM and 512 bytes of EEPROM. Fuzzy logic kernels have been written for the Neuron® chip that can be used with support tools to easily implement fuzzy logic as part of the control portion of the distributed network [25].

6.8.4 MI-Bus

Motorola has developed a protocol called the MI-Bus for a low-cost network solution. This approach for class A operation is built around Motorola's HC05 and HC11 MCUs and is designed to achieve low cost in certain master-slave control applications. The MI-Bus protocol has been designed for direct control of (low frequency) power loads [26]. The master-slave arrangement allows one master to control as many as eight slaves. The MI-Bus can be driven out of the I/O port. It should be noted that even with the considerable effort that has been expended on J1850 and CAN specifications and hardware implementations that have been developed, a need for an alternative solution can be identified by a customer with sufficient volume to justify the development of a custom solution.

The MCU initiates the protocol by issuing a sequence of 8 data bits that contain the 3-bit address and 5-bit control. The three address bits allow up to eight devices to be addressed. After transmitting data, the MCU reads serial data from the selected device. Bus access method, frame sequence protocol, message validation, error detection, and default values are all defined for the MI-Bus.

6.8.5 Other MCUs and Protocols

The Profibus protocol is available in RAM microcode in a number of semiconductor devices. One single-chip local area network (LAN) solution has been developed to interface Ethernet and a variety of different sensor networks, including Profibus [27]. As shown in Figure 6.12, network nodes can be connected to an Ethernet hub using twisted-pair cables. The MC68360 in this application has a 32-bit CPU that provides 4.5 MIPS at 25 MHz for power and flexibility in any existing Ethernet system. Because Ethernet is the leading number of installed LANs, interoperability is essential.

6.9 Other Aspects of Network Communications

In addition to the protocols already discussed, MCU manufacturers use serial communications protocols between ICs that are on the same printed circuit board. Those interfaces can also be considered for subsystems in communications networks. When MCU protocols or any of the previously discussed

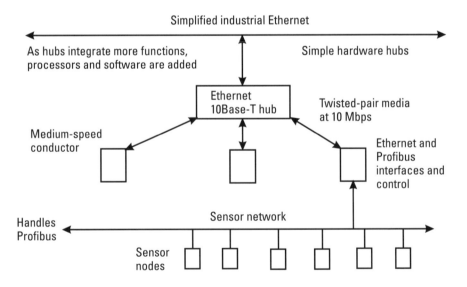

Figure 6.12 Ethernet and Profibus interface.

protocols are used, the need to transition between various protocols can occur, especially in complex systems. MCU protocols, the transition between protocols, and modular protocol solutions are the final topics on communication in this chapter.

6.9.1 MCU Protocols

The serial peripheral interface (SPI) is a high-speed protocol that is used for synchronously transferring data between master and slave units in an onboard serial network. As shown in Figure 6.13, the SPI module is a defined building block that can be used for transmitting digital information in a system through a software-defined protocol. This module is already contained in a number of MCU families. The queued SPI (QSPI) is an intelligent, synchronous serial interface with a 16-entry, full-duplex queue. It can continuously scan up to 16 independent peripherals and maintain a queue of the most recently acquired information without CPU intervention.

The serial communications interface (SCI) transmits using only two pins on MCUs with an SCI peripheral. The SCI protocol was used to develop the MI-Bus (discussed in Section 6.8.4). That option is discussed in more detail in Section 6.9.2.

The inter-IC, or I²C, is a two-wire half-duplex serial interface with the data transmitted/received by the most significant bit first. The two wires are a serial data line (SDL) and a serial clock line (SCL). The protocol consists of a start condition, a slave address, *n* bytes of data, and a stop condition. Each byte is followed by an acknowledge bit. The I²C peripheral can be interfaced by means of a synchronous serial I/O port (SIOP).

6.9.2 Transition Between Protocols

Because several communication protocols exist, the transition between protocols frequently is required. A gateway node provides the transition between

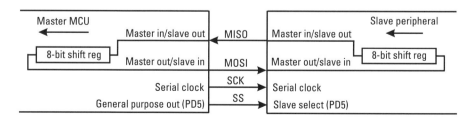

Figure 6.13 The SPI.

networks that use different protocols. In some cases, custom hardware and software gateways convert information from one system's language to that of the other system. Ideally, the MCU in this node has one or both protocols that allow an easy transition. In the worst case, the MCU uses a separate peripheral for each protocol. Figure 6.14 shows the gateway master from the CAN to an MI-Bus. On the master side, the SCI protocol common to many MCUs can be used. On the other side of the gateway, the MCU has a CAN module for communication on the CAN network. That allows control information to pass from the CAN network across to drive the output loads. On the slave side, the I/O controller provides the interface to the load. A stepper motor controller, the MC33192, was the first device to control loads in an automotive environment using the MI-Bus [26]. The integrated device has both the MI-Bus controller and a dual full-bridge driver for the stepper motor. Other modules can easily be designed to control other loads by simply revising the output portion of the MI-Bus device.

6.9.3 Transition Between Systems

A gateway can provide security and safety for control systems. Modern vehicles provide a unique environment for communications, computer, and

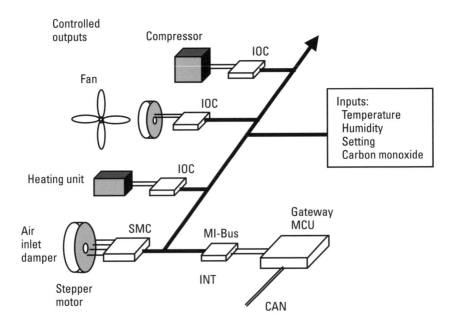

Figure 6.14 CAN to MI-Bus gateway and various sensors and loads.

entertainment products. In addition, intelligent transportation system (ITS) efforts to improve the efficiency and safety of surface transportation will require new technology in several consumer-oriented products to achieve their objectives. However, auto manufacturers are concerned that the reliability and functionality of existing systems may be compromised by the addition of hardware and software that they did not develop or qualify for use in vehicles. As a result, the SAE has developed the ITS databus (IDB) standard. The IDB really acts as a firewall, controlling and limiting the flow of information between the two networks to only that information that is valuable to both. The ITS databus approach uses a gateway MCU to interface a network of consumer electronics components and systems with existing vehicle systems as shown in Figure 6.15 [28].

6.9.4 The Protocol as a Module

The protocol can be easily reconfigured into a new MCU design using the modular design methodology described in Chapter 5. In addition to the CPU and other modules, serial communications protocols, including CAN, both versions of SAE J1850, SPI, SCI, and I²C can be treated as a modular peripheral.

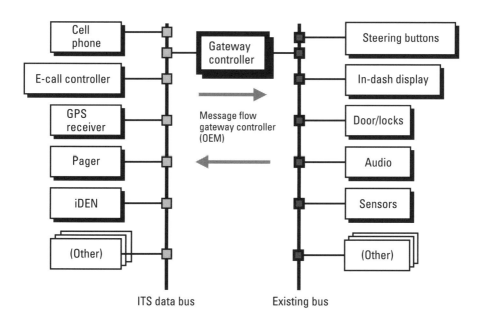

Figure 6.15 The IDB, which isolates consumer-added products from the manufacturer-installed systems in the vehicle.

That allows new features to be easily added to an MCU with an integrated proto-col as required for different applications. Also, a protocol can be easily added to an existing MCU design that will become part of a communication network.

6.10 Summary

The ability of sensors to communicate with other portions of the control sys-tem will allow more intelligence at the sensor node and more distributed con-trol. Protocols that have available silicon solutions have a higher likelihood of acceptance. Many types of standards have been developed; however, implemen-tation options may limit their compatibility. An entirely new protocol devel-oped for sensor-specific applications would have to be well defined and accepted by sensor users to justify the development cost necessary to implement the protocol in silicon.

References

[1] Pullen, D., "Smart Sensors: Adding Intelligence to Transducers," *Proc. Sensors Expo West*, San Jose, CA, Mar. 2–4, 1993, pp. 117–136.

[2] Raji, R. S., "Smart Networks for Control," *IEEE Spectrum*, June 1994, pp. 49–55.

[3] Madan, P., "LonWorks™ Technology for Interfacing and Networking Sensors," *Proc. Sensors Expo*, Philadelphia, Oct. 26–28, 1993, pp. 225–242.

[4] Najafi, N., and K. D. Wise, "An Organization and Interface for Sensor-Driven Semicon-ductor Process Control Systems," *IEEE Trans. Semiconductor Manufacturing*, Vol. 3, No. 4, Nov. 1990, pp. 230–238.

[5] Bredius, M., F. R. Riedijk, and J. H. Huijsing, "The Integrated Smart Sensor (IS^2) Bus," *Proc. Sensors Expo*, Philadelphia, Oct. 26–28, 1993, pp. 243–247.

[6] Meisterfeld, F., "Chapter 26: Multiplex Wiring Systems," *Automotive Electronics Hand-book*, 2nd ed., R. K. Jurgen (ed.), New York: McGraw-Hill, 1999, pp. 26.1–26.76.

[7] Frank, R., and C. Powers, "Microcontrollers With an Integrated CAN Protocol," *Proc. Power Conversion and Motion*, Sept. 17–22, 1994, pp. 1–10.

[8] Terry, K., "Software Driver Routines for the Motorola MC68HC05 CAN Module," Motorola AN464, 1993.

[9] Lawrenz, W., and M. Pauwels, "OSEK/VDX—The Operating System for Distributed Systems in Cars," *Proc. Convergence 96*, Dearborn, MI, Oct. 21–23, 1996, pp. 281–285.

[10] Chatha, A., "Fieldbus: The Foundation for Field Control Systems," *Control Engineering*, May 1994, pp. 77–80.

[11] *Neuron Chips*, Motorola Brochure BR1134/D Rev. 1, Sept. 1993.

[12] Johnson, D., "Looking Over the Bus Systems," *Control Engineering*, Dec. 1995, pp. 56–64.

[13] McMillan, A., "Fieldbus Interfaces for Smart Sensors," *Proc. First IEEE/NIST Smart Sensor Interface Standard Workshop*, Gaithersburg, MD, Mar. 31–Apr. 1, 1994, pp. 133–148.

[14] "Bus of the Month ARCNet," *Control Engineering*, Oct. 1997, pp. 101–102.

[15] Sack, T., "Building Buses: The Plot Thickens," *Electrical Review* Vol. 224, No. 20, Oct. 1991, pp. 28–29.

[16] Jancsurak, J., "Electronics Breakthroughs Now and On the Horizon," *Appliance Manufacturer*, May 1993, pp. 29–30.

[17] Strassberg, D., "Home-Automation Buses: Protocols Really Hit Home," *EDN*, Apr. 13, 1995, pp. 69–80.

[18] MC68705V8 Motorola Specification Rev. 2, 1993.

[19] Powers, C., "Example Software Routines for the Message Data Link Controller Module on the MC68HC705V8," Motorola AN1224, 1993.

[20] Powers, C., "J1850 Multiplex Bus Communications Using the MC68HC705C8 and the SC371016 J1850 Communications Interface (JCI)," Motorola AN1212, 1992.

[21] CAN Module, Motorola Product Preview, 1990.

[22] Cole, B., "Motorola Tunes PowerPC for Auto Applications," CMP TechWeb (CMPT), Apr. 21, 1998..

[23] MC68HC05X4/MC68HC705X4 Advanced Information, Motorola, 1993.

[24] Neumann, K.-T., "The Application of Multiplex Technology," *Automotive Technology Internat'l '95*, pp. 199–205.

[25] "Fuzzy Logic and the Neuron Chip," Motorola AN1225, 1993.

[26] Burri, M., and P. Renard, "The MI-BUS and Product Family for Multiplexing Systems," Motorola EB409, 1992.

[27] O'Dell, R., "Ethernet Joins the Factory Workforce," *Machine Design*, July 11, 1994, pp. 78–82.

[28] Powers, C., and R. Frank, "The Consumerization of the Automotive Environment: The ITS Data Bus," SAE Future Transportation Technologies Conference, SAE 972839, San Diego.

7

Control Techniques

A machine is intelligent when it does the things of a common three year old.
—Marvin Minsky, creator of the concept of artificial intelligence

7.1 Introduction

Distributed control systems (DCS) have brought decision making to the sensor. The built-in intelligence for the smart sensor can take advantage of a variety of existing and developing control techniques. The extent to which new computing paradigms will affect sensors depends on the need for control at the sensor and the cost effectiveness of using those new techniques. Existing control techniques have grown from programmable logic controllers (PLCs) and PC-based instruments. Proportional-integral-derivative (PID) control and the state machine are the basis of many control systems. However, the newest artificial intelligence approaches of fuzzy logic and neural networks will increasingly affect sensor-based systems. Fuzzy logic is already finding several applications in appliances and automobiles. Adaptive control techniques using the higher performing MCUs and DSPs will allow sensors to adjust to aging effects during their operating life. Control strategies that implement computationally intense algorithms will be able to shrink a proof-of-concept laboratory demonstration into a useful solution that can have popular applications based on the increased computing capability of MCUs and DSPs.

7.1.1 Programmable Logic Controllers

PLCs are used in industrial applications to control a variety of logic and sequencing processes. A typical PLC consists of a CPU, a programmable memory, I/O interfaces, and a power supply [1]. Smaller, less expensive PLCs can be used and the response time and interference from noise reduced by "combining a sensor and microprocessor into one unit, commonly called a smart sensor" [1]. Functions performed by smart sensors in that role include correcting for environmental conditions, performing diagnostic functions, and making decisions.

7.1.2 Open- Versus Closed-Loop Systems

The simplest control system is an open-loop system. The sensor input goes to a processing unit, which produces an output. As shown in Table 7.1, that requires basic mechanics and physics knowledge, as well as mechanical-to-electrical interface knowledge for a simple system such as a fuel indicator [2]. The table shows the increasing knowledge level required to implement increasingly more complex systems with automotive applications as examples.

A closed-loop system can use the sensor to modify the control strategy and use a control strategy to improve the performance of the sensor [3]. Figure 7.1 is a comparison of a capacitive surface-micromachined accelerometer using an open-loop signal conditioning circuit and a closed-loop signal conditioning circuit. The closed-loop system has advantages over the open-loop system, including improved linearity and wider dynamic range. Table 7.2 is a comparison of several parameters [3].

In general, a single-loop closed-loop system has a comparison node for a set point versus an input (the feedback signal) from a sensor regarding the status of the process that is being controlled. On the basis of the comparison, control action is taken to increase, reduce, or maintain speed, flow, or some other variable controlled by a mechanical actuator such as a valve or motor. The PID control algorithm is a common technique in a closed-loop system [4].

7.1.3 PID Control

Most single-loop controllers include at least three principal control actions: proportional, integral (or reset), and derivative (or rate). Each action corresponds to a specific technique for controlling the process set point. Proportional adjustments amplify the error by a constant amount relative to the size and sign of the deviation. The proportional band has a range of deviations in

Table 7.1
Types of Vehicle Control Systems (*After:* [2])

System	Knowledge/Skill	Application Example
Open loop	Basic mechanics, physics; mechanical-electrical interface	Fuel indicator
Closed loop	Basic feedback theory	Automatic temperature control
Complex processor	Multiple function; sequential processes	Engine management
Interactive system	Interactive system controls; control analysis CAE	Antilock brakes and traction control
Total vehicle control	Architecture tools; failure mode management; communication theory; structured software	Intelligent vehicle
Global	Higher level languages; simulators	External environment inputs

Table 7.2
Comparison of Open- and Closed-Loop Systems

Parameter	Open-Loop System	Closed-Loop System
Latchup	Possible	Not possible
Mechanical fatigue	Possible	Minimized
Linearity	Medium	High
Dynamic range	Narrow	Wide
Frequency response	Set by damping	Set by electronics
Sensitivity	Set by mechanical design	Set by electronics

the percentage of full scale that corresponds to the full operating range of the control element [4].

Integral control is a function of time, unlike proportional control, which has no time associated with it. Integral control responds to the continued existence of a deviation such as an offset. The combination of proportional and integral actions results in a change at a steady rate for each value of deviation. The integral rate is usually measured in repeats per minute.

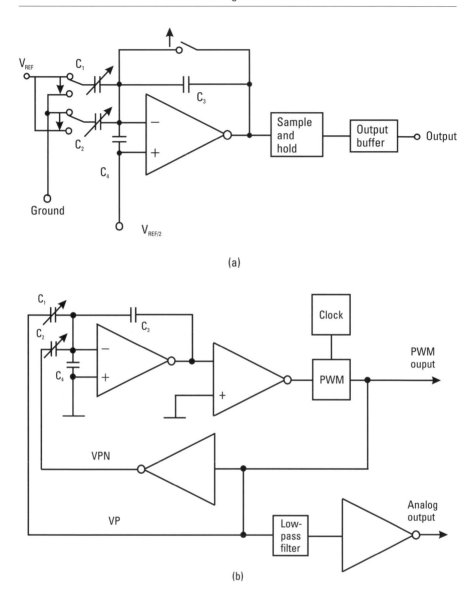

Figure 7.1 (a) Open-loop versus (b) closed-loop accelerometer circuit.

Derivative control is used only in combination with the proportional or integral control action. Changes in speed or direction of the deviation initiate derivative control action. That action reduces the time that the system would otherwise take to reach a new set point determined by the proportional control. Controllers with derivative control have rate amplitude and rate time adjustments. PID control can be implemented in both analog and digital systems [4].

The PID control algorithm is defined by

$$T(s) = C(s) / R(s) = 1 \qquad (7.1)$$

where: $T(s)$ is the desired transfer function, and $R(s)$ and $C(s)$ are Laplace transforms of the reference input $r(t)$ and the controlled output $c(t)$ [5].

If $C(s)$ is not measured, the system is open loop. Otherwise, a measurement of $C(s)$ is used to provide closed-loop feedback control. Figure 7.2 shows the location of those parameters in the analog control system [5]. For $r(t)$ varying with time, a tracking problem must be solved. For $r(t) = constant$, the problem is one of regulation. Disturbances in the forward transmission path and the feedback path can prevent $c(t)$ from ideally following $r(t)$.

The process being controlled typically dictates if only proportional, proportional and integral, or full PID should be implemented. As in most systems, the most appropriate solution is the simplest one that will achieve the desired results. Control systems based on digital techniques can improve previous analog systems. Figure 7.3 shows the simplification that results when a single rotary encoder provides position, speed, and commutation data in a speed-control system [6]. The number of connections to the motor is reduced from 13 to 8. The speed computation block is used to interpolate the encoder output signal for improved resolution.

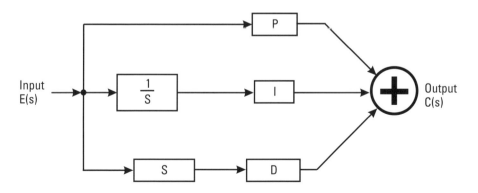

Figure 7.2 PID control.

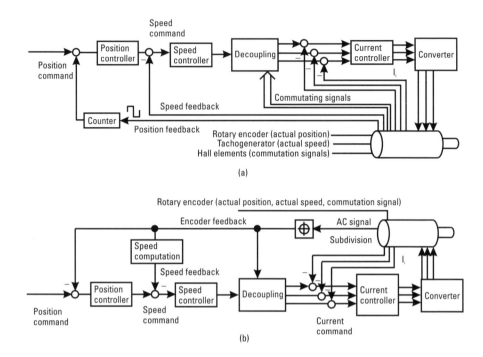

Figure 7.3 (a) Analog versus (b) digital speed control. (*After:* [6].)

7.2 State Machines

The state machine is one of the most commonly implemented functions in programmable logic [7]. The state machine is actually a sequencing algorithm. It allows an interrupt service routine to execute an algorithm that extends over many interrupts. State machines can be open loop or closed loop [8]. State machines are developed from a state diagram and a state table. The state diagram is determined at the beginning of the design process from a description of the problem. A natural language describes the intended circuit function. A key area of interest is the synthesis of the state machine, a process that will allow rapid and easy development of the circuit from a functional description of the state machine. If only a small subset of inputs is sensed while the circuit is in any given state, then an algorithmic state machine is used in place of a state diagram.

A finite-state machine is a mathematical representation of a digital sequential circuit in which abstract symbols, not binary codes, represent the states. Finite-state machines can be specified in state transition tables or circuits consisting of logic gates and flip-flops. These dedicated controllers can be either synchronous or asynchronous and can be implemented in hardware or software.

7.3   Fuzzy Logic

Fuzzy logic control theory is an alternative to lookup tables and algorithm calculations that is easier to develop and has performance advantages in many applications. Fuzzy logic can be defined as a branch of logic that uses degrees of membership in sets rather than a strict true/false membership [9]. Fuzzy logic is finding acceptance in consumer, automotive, industrial, and even decision-making applications [10]. Table 7.3 gives examples of existing applications of fuzzy logic.

In contrast to PID control, which models the system or process being controlled, fuzzy logic focuses on the human operator's behavior [11]. Fuzzy logic defines a partial set membership, not the strict one normally associated with the Boolean logic used in MCUs and DSPs. Rules stated in natural language establish membership to the fuzzy sets.

Figure 7.4 shows an example of fuzzy logic membership for a pressure (or vacuum) input. The fuzzy set allows membership with different grades, each expressed by a number in the interval [0, 1]. In that example, membership functions and rules are established for the five pressure ranges of very low, low, normal, high, and very high. The membership value for intermediate locations is some number between 0 and 1. A rule for the very high pressure could be this: *If* the pressure is very high, *then* decrease the power greatly, with a weight

Table 7.3
Applications of Fuzzy Logic

Consumer	Automotive	Industrial	Decision Support
VCRs	Air conditioning	Extruding	Building HVAC
Camcorders	Transmission control	Mixing	Medical diagnostics
Televisions	Suspension control	Furnace controls	Transportation
Vacuum cleaners	Cruise control	Temperature controls	Elevator controls
Clothes dryers		Conveyor controls	
Washing machines			
Hot water heaters			

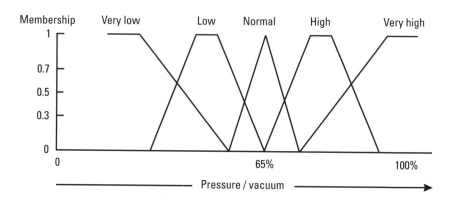

Figure 7.4 Fuzzy logic membership functions for pressure control.

of 0.7 assigned for "greatly." For low pressure, the rule could be: *If* the pressure is low, *then* increase the power slightly, with a weight of 0.3 assigned for "slightly."

The final part of the fuzzy logic process is defuzzification. The defuzzification process takes a weighted average to translate the fuzzy outputs into a single crisp value. Output membership functions are typically defined by singletons. A singleton in an 8-bit system is defined as an 8-bit value that represents the output value corresponding to the linguistic label of the output system.

To simplify the implementation of fuzzy logic on existing MCUs, a fuzzy kernel can be developed by the MCU manufacturer [12]. A fuzzy kernel, or engine, is software that performs the three basic steps of fuzzy logic: fuzzification, rule evaluation, and defuzzification. As shown in Figure 7.5, the fuzzy-inference unit (FIU) receives system inputs and information from the knowledge base for each step of the process. All application-specific information is contained in the knowledge base that is developed independently from the fuzzy-inference program.

With fuzzy logic, applications that were once thought to be too complex to be practical are being easily controlled. Furthermore, the rules in a fuzzy system often hold true even if the operating parameters of the system change. That typically is not true in conventional control systems.

In general, fuzzy logic does not require special hardware. The software requirements can be low based on a limited number of rules to describe the system. For most control tasks, software running on standard processors can perform fuzzy logical operations. Specialized hardware does simplify code development and boost computational performance for low-cost applications.

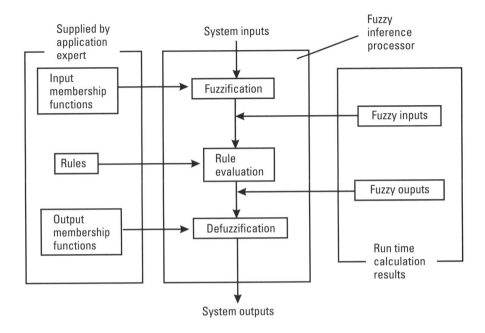

Figure 7.5 FIU.

Fuzzy logic allows 8-bit MCUs to perform tasks that otherwise may have required a more powerful MCU, such as a RISC processor.

Fuzzy logic has been used with multiple sensors to produce a mobile security robot for evaluating the environment in an art gallery [13]. As shown in Figure 7.6, the fuzzy logic certainty engine in the robot weighs multiple variables, creates a composite picture from data supplied by the sensors, and performs real-time threat identification and assessment. The robot is able to analyze data based on their location, the time of day, historical trends, and external conditions. A wide range of safety threats detrimental to art objects has been identified by the robot during a 36-month evaluation. An obvious extension of this technology is air quality assessment for conditions harmful to humans.

7.4 Neural Networks

Unlike fuzzy logic, neural networks allow the system instead of an expert to define the rules. By definition, a neural network is a collection of independent processing nodes that solve problems by communicating with one another in a

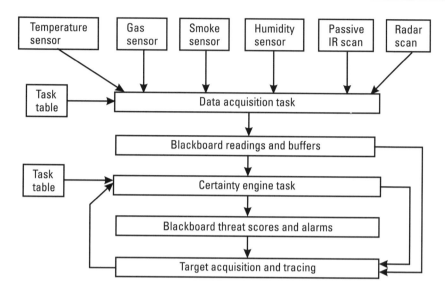

Figure 7.6 A variety of sensors providing input to the fuzzy logic certainty engine of a mobile robot. (*After:* [13].)

manner roughly analogous to neurons in the human brain [11]. Neural networks are useful in systems that are difficult to define. They have the additional advantage of being able to operate in a high-noise environment. Complex or numerous input patterns are among the problems that neural networks are addressing [14].

Figure 7.7(a) shows the structure of a neuron [14]. The neuron learns from a set of training data. Each input to the neuron is multiplied by the synapse weight. The neuron sums the results of all the weighted inputs and processes the results with a typically nonlinear transfer function to determine the output. A local memory stores previous computations and modifies the weights as it learns.

A neural network consists of a number of neurons connected in some manner, as indicated in Figure 7.7(b). In that model, a hidden layer exists that prevents direct interaction between the input and the output. However, the hidden layer is not required, and more than one may exist. Also, the number of neurons does not have to be the same in each layer. To train or condition the neuron for proper response, several sets of inputs with known results are applied to the neural network. The output is compared for each set of inputs to the known results, and the weights are adjusted to compensate for errors.

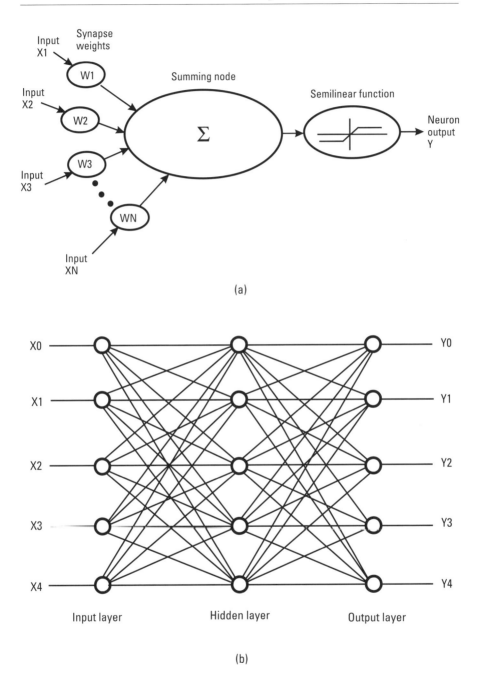

(a)

(b)

Figure 7.7 (a) Neuron and (b) neural network. (*After:* [14].)

An example of a potential application for a neural network is the control of an automotive fuel injection system [15]. Production vehicles meet current emissions regulations using calibration and lookup tables. However, emission regulations for 2003 have further reductions of hydrocarbons (HC), oxides of nitrogen (NO_X), and carbon monoxide (CO) emissions that may require the function-approximation, learning, and adaptive capabilities of neural networks. By using a neural network control, a stoichiometric air-fuel ratio (A/F) can be maintained over the life of the vehicle even if the engine dynamics change. Experimental results with a neural network and a linear A/F sensor have demonstrated the capability to control stoichiometry within ±1%, which was better than the production control unit.

7.5 Combined Fuzzy Logic and Neural Networks

Fuzzy logic and neural networks are being combined to utilize the best features of each technology. One approach begins with a set of fuzzy rules that have been well tuned by an expert using trial-and-error methods. A neural-like adaptive mechanism is then installed in the fuzzy system to handle exceptional circumstances after the system is in use. These systems compensate for load variables and for wear that occurs over time [16].

In an alternate approach, the fuzzy system is coarsely defined by experts. The fuzzy rule base is then refined with a neural network. The neural network adapts to minimize errors.

Another approach combines fuzzy associative memories with neural networks. In general, an associative memory is a neural architecture used in pattern recognition applications. The network associates data patterns with specific classes or categories it has learned. That combination produces a system in which the neural network front end learns rules from training data and then supplies those rules to a fuzzy logic back end to execute the rules [16].

The previous examples started with a fuzzy system and applied neural network learning. Other researchers have started with neural networks and applied fuzzy logic. In those systems, the network adapts in a more intelligent manner. The ultimate technology in this area could be evolutive learning, the application of genetic algorithms to combine fuzzy and neural systems automatically. Genetic algorithms are guided stochastic search techniques that utilize the basic principles of natural selection to optimize a given function [17]. These advanced control concepts are being explored for process optimization in the manufacture of complex ICs.

7.6 Adaptive Control

Adaptive control is a system of advanced process control techniques that is capable of automatically adjusting or adapting to meet a desired output despite shifting control objectives and process conditions or unmodeled uncertainties in process dynamics [17]. Because of their capacity to learn, neural networks have the capability to provide adaptive control, but they are not the only means. As shown in Figure 7.8, the difference between the desired performance and the measured performance in the control system is the starting point for the adaptive process [18]. The adaptive system modifies the parameters of the adaptive controller to maintain the performance rating close to the desired value.

All adaptive control systems have a conventional servo-type feedback loop and an additional loop designed to identify the online process and determine the parameters to be adjusted on the basis of the process parameters. The addition of an identification algorithm and an outer control loop maintains performance in the presence of disturbances. An indirect adaptive scheme has an additional outer loop to perform system identification, and it provides access to process parameters for monitoring and diagnostics [18].

Adaptive control can be obtained in fuzzy logic systems using a controller that operates at a higher level to perform tuning and adaptive control functions [19]. Figure 7.9 compares the architecture for that approach to a low-level direct closed-loop system. Adjustments that the fuzzy logic controller performs

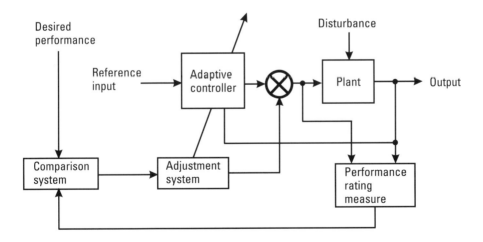

Figure 7.8 Adaptive control system.

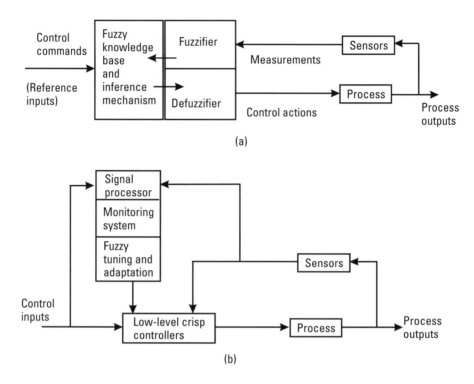

Figure 7.9 (a) Low-level direct control and (b) high-level tuning/adaptive control in fuzzy logic control system. (*After:* [19].)

include online and offline tuning of the controller parameters, online adaptation of the low-level controller, and self-organization and dynamic restructuring of the control system.

7.6.1 Observers for Sensing

Observers, or algorithms that correct for variations from physical models, are increasingly being used in more complex systems such as motor controls and automotive engine controls. A discrete time domain representation of an observer in a motor control system is shown in Figure 7.10 [20]. In Figure 7.10, the intent is to compute the best-estimated value for velocity from the measured position of the rotor. In Figure 7.10, u is the input voltage, A is a matrix representing the motor's dynamics, B is the input vector, D is a vector that shows which of the system variables is the output (position), X is the state-variable vector that includes two components: position and velocity, Z is the

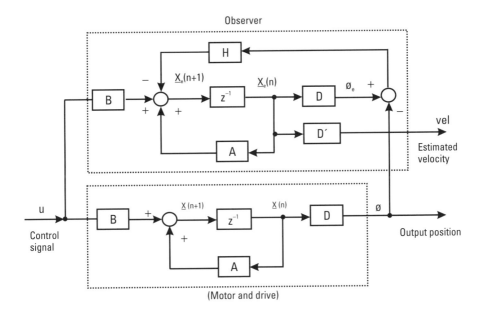

Figure 7.10 The block diagram of an observer for closed-loop velocity control.

delay operator, and *H* is the observer vector. The lower portion of Figure 7.10 shows the block diagram of the actual motor, and the upper portion contains the observer algorithm that is in the software [20].

The observer estimates both velocity and position. A comparison of the estimated position with the measured position is used to correct for the next sampling period. Estimation error is used to achieve a better estimate for the next sampling interval. The observer gain vector *H* is used to modify the observer dynamics.

Feedforward techniques use sensors to modify the estimated values for system inputs that are in the observer model [21]. The growing complexity of automotive engine controls to meet increasingly stringent emissions targets may require systems that anticipate, act, and then adjust algorithms based on measurements on a cylinder-by-cylinder basis. Model-based methods can also reduce the calibration effort that takes considerable time and effort for each internal combustion engine that a manufacturer produces. Sensor-based intelligent decision making has also been investigated as a possible solution in computer-based untended machining [22]. Those and other techniques that expand the capability of neural networks rely heavily on sensor-system synergy.

7.7 Other Control Areas

Computationally intense models such as hidden Markov modeling (HMM) (which is used in automatic speech recognition) will be used more frequently as available computing power increases [23]. Higher performance for real-time control is achieved in DSPs and multiprocessor systems. For example, automatic speech recognition (ASR) has been implemented in ROM on a DSP chip. The HMM technique treats speech as a stochastic process for matching the input to a word command. In those systems, sensor input from a microphone directly affects the signal-to-noise ratio of the input. Limited bandwidth and high-noise systems, such as telephone lines, still cause problems to some existing speech recognition systems. However, continuing improvements in algorithms and the computational engine are improving the process.

Multiprocessor computing is used today for controlling industrial robotics. Those powerful processors support the implementation of machine vision, speech recognition, and optical character recognition. Techniques used in multiprocessing include common Fourier transform, convolution, linear prediction, dynamic time warping, template matching, and spectral processing [24].

The discrete Fourier transform is used to determine the frequency components of a time series. Multiplying a matrix of cosine and sine functions with the time-series vector yields the frequency-series vector. By computing a 64-element time series vector concurrently on 64 processors (one component per processor), processing time was reduced to $\frac{1}{64}$ of the uniprocessor's time.

Convolution is used for filtering a time series. It requires calculating the inner products of time-series sections with a coefficient vector that defines the filter characteristics. By using a 16-element convolution kernel or software function and distributing the elements in a 128 processor with only one broadcast of the time-series vector, the computation was performed 128 times faster.

Linear prediction is used to obtain smoothed, compact spectral representations of a signal. It requires calculating the autocorrelation function for the time signal. That is identical to convolution calculation with the convolution kernel replaced by a section of the time series for autocorrelation. By processing one coefficient per processor, the autocorrelation function is calculated N times faster, where N is the number of autocorrelation coefficients computed. Linear predictive coding is frequently used for speech recognition.

Dynamic time warping calculates the similarity between two signals that have differing nonlinear time-alignment patterns. The computation determines the optimum path through a matrix of distances formed from the squared Euclidean distances between discrete sample sections of the signals. Multiple paths are explored based on the selected algorithm, and the minimum total distance defines the similarity and best alignment path. Template matching in linear-vector quantization classifier requires a similar calculation of the Euclidean distance between an unknown vector of features and a lexicon of templates. Time warping is another approach for speech recognition.

Spectral processing involves a sequence of computations. The computations include a weight vector, Fourier transform, conversion from real to imaginary components to power, and conversion from power to decibels. The final calculation of the power components of the frequency spectrum is converted to decibels using a standard linear-to-logarithmic library function. The amount of processing is indicative of speech and image recognition.

Associative memory is used in pattern recognition applications, in which the network is used to associate data patterns with specific classes or categories it has already learned [17].

The control areas discussed in this section are aimed at some of the high-end computing requirements in sensory systems. As shown in Table 7.4, these technologies are essential for many smart-sensing applications. More than one approach can be used to implement a solution for these complex systems.

7.7.1 RISC Versus CISC

The complex instruction set computer (CISC) architecture is the established approach for desktop computing and embedded control applications. An

Table 7.4
Emerging Techniques for Smart Sensing Applications

Technique	Enabling Technology
Voice recognition	DSP, neural network (NN), RISC, multiprocessing (MP)
Handwriting recognition	DSP, MP, NN, fuzzy logic (FL)
Multimedia	DSP, FL, RISC, MP
Digital sound	DSP
FAX/modem	Radio frequency integrated circuits (RFICs), CISC
Data storage	Flash, DRAM, SRAM
CD-ROM	CISC
Virtual reality	DSP, RISC, FL, MP
Synthesized speech	DSP, RISC, NN, MP
Data compression	DSP
Pattern recognition	DSP, high-performance MCU, NN

embedded system has one or more computational devices (which may be MPUs or MCUs) that are not directly accessible to the user of the system. A relatively new addition for computing requirements is the reduced instruction set computer (RISC) architecture. Computer and workstation manufacturers seeking the highest performance levels are using RISC architecture. Embedded computing applications are also utilizing the performance that this approach provides. The transition from CISC to RISC can be understood by analyzing the following equation for MPU performance:

$$Performance = \frac{MHz \cdot Instructions \ / \ Clocks}{Instructions} \qquad (7.2)$$

CISC-based architectures allocate transistors to decreasing the denominator of (7.2) [25]. However, based on recent improvements in the price performance of semiconductor memory and significant advances in compiler technology, transistors can be reallocated to increasing the numerator. That is the basis of the RISC architecture that frees up transistors for integrating additional components. The additional components can be peripherals or microcoded kernels that improve the performance in a specific application.

Two approaches, CISC and RISC, are now available to solve a variety of application problems in several markets. Because no single computing solution is correct for all applications, designers must choose the approach that satisfies their design criteria. For a given application, there is a best-fit solution and more than one feasible solution. CISC-based designs will continue to satisfy a significant number of applications, especially in embedded control. However, RISC has established a credible presence in a range of commercial computing environments and its use will expand in embedded applications. RISC architecture in an embedded environment does not take advantage of the 64-bit bus, zero wait state memory, or the amount of memory integrated on one desktop MPU chip. To be affordable, the RISC performance is compromised but still beyond the capabilities of today's CISC MCU.

CISC MCUs range from simple 4-bit units and the most popular 8-bit units up to 32-bit units. The lower cost versions are easily combined with single sensors such as a pressure, flow, or acceleration sensor to provide a smart sensor. The high-performance 32-bit CISC and RISC units are allowing complex strategies and diagnostics in automotive engine and powertrain control systems. They also are required for more complex sensing in imaging and navigation systems.

7.7.2 Combined CISC, RISC, and DSP

An interesting demonstration of the value of both of CISC and RISC is provided in a communications control circuit is shown in Figure 7.11 [26]. This is a single-chip solution incorporating CISC, RISC, *and* DSP technology. CISC technology (68000-based) is used for the integrated multiprotocol processor. Three serial communications channels handle high-level data-link control (HDLC), asynchronous (UART), BYSYNC, and synchronous protocols. A Personal Computer Memory Card International Association (PCMCIA) block provides a direct interface to a PCMCIA bus (PC card). RISC is used for the communication processor. As a result, the ports operate independent of the 68000-host processor and allow the 68000 to handle higher level control tasks [27]. DSP technology provides increased performance through the parallel (Harvard) architecture. That allows it to simultaneously fetch program, X-data, and Y-data memory. Because both cores use common memory, cost is reduced.

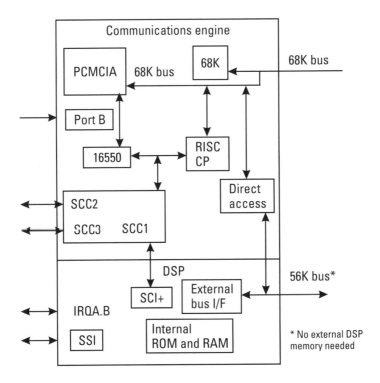

Figure 7.11 Block diagram of combined CISC, RISC, and DSP for data communications.

The communication processor can be applied to sensing applications that include voice compression in tape answering machines and simultaneous voice/data transmission. It demonstrates the extent that technologies can be integrated and indicates the potential for sensor control integration in the future.

7.8 The Impact of Artificial Intelligence

Fuzzy logic and neural networks will undoubtedly affect not only the control systems, but also the sensors that are used in these systems. One analysis shown in Figure 7.12 suggests that artificial intelligence will require sensors with "lower" accuracy and, subsequently, will cost less [28]. At the same time the number of sensor applications will increase as a result of artificial intelligence (and the lower cost). Artificial intelligence includes the newer control techniques, such as fuzzy logic and neural networks.

The control decisions made in fuzzy logic systems can be made in spite of absolute accuracy from the sensory data [29]. As a result, performance specifications will change from point-to-point to domain-type specifications.

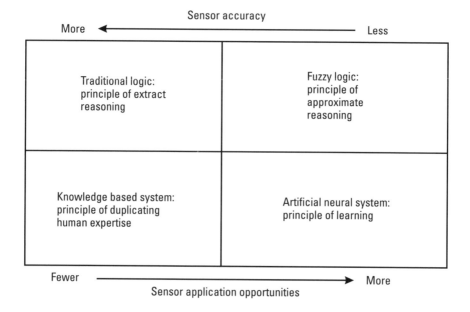

Figure 7.12 Impact of artificial intelligence on sensors. (*After:* [28].)

Domain-type specifications allow the sensor to function with wider tolerance in terms of repeatability, accuracy drift, and linearity. A natural extension of the fuzzy logic rules to the sensor has the sensor expressing the measurement in terms of grade of membership to the predefined domain. Those principles have already been applied to color, proximity, fluid control, and position sensors.

An extensive analysis of classic control theory, fuzzy logic, neural networks, and rough set theory has been performed for smart sensors [30]. The authors have defined an operator's inference model that allows evaluation of the situation, assignment of the situation to a characteristic state, and selection and execution of the proper characteristic control. The model has been used to evaluate smart sensors and control system design. The rough set in the analysis is similar to fuzzy logic, except that a rough ADC replaces the fuzzifier in the controller, and a rough DAC replaces the defuzzifier. The inference engine in the controller evaluates the decision table for the rough set. The decision tables used to derive the rules in the rough controller are simple to understand and easy to edit.

As part of their findings, the authors determined that classic control theory, a mature technology with widely supported hardware, can accomplish more complex computing by increasing the millions of instructions per second. Fuzzy logic has commercial developments available that are making its acceptance widespread, but it does have safety and reliability disadvantages. The rough set can be used for pattern matching and is fast and low cost, but it does have stability and completeness concerns similar to those with fuzzy logic and neural networks. Neural networks can possibly provide an all-analog system that is easily integrated with sensors, but they have the added problem of an unknown decision basis. Attention to the areas of concern, however, is allowing all those systems to be evaluated for new applications. The combination of those control techniques with sensors will achieve new levels of smart sensing.

7.9 Summary

This chapter discussed control techniques and significantly improved computing capabilities. Fuzzy logic is already finding broad acceptance in many control applications. More complex systems will benefit from fuzzy logic and other advanced approaches, such as model-based algorithms. Some of those systems will use more feedback and, consequently, more sensors. However, the sensors, at least in some instances, may not have to be as accurate and therefore could offset at least some of the added cost of getting smarter.

References

[1] "Controls and Sensors," *Power Transmission Design*, Jan. 1995, A179–A190.

[2] Gormley, J., and D. A. MacIsaac, "The Systems Approach (Better Late Than Not at All)," *Proc. Convergence '88*, Dearborn, MI, Oct. 17–18, 1988, pp. 107–118.

[3] Dunn, W., and R. Frank, "Automotive Silicon Sensor Integration," SAE SP-903 *Sensors and Actuators 1992*, SAE Internat'l Congress & Exposition, Detroit, MI, Feb. 24–28, 1992, pp. 1–6.

[4] Wachstetter, J., "Understanding the Principal Control Actions of Single-Loop and Multi-loop Controllers," *I&CS*, Aug. 1994, pp. 45–51.

[5] Stokes, J., and G. R. L. Sohie, "Implementation of PID Controllers on the Motorola DSP56000/DSP56001," *Motorola APR5/D Rev. 1*, 1993.

[6] Hagl, R., and S. Bielski, "Rotary Encoders Make Digital Drives Dynamic," *Machine Design*, Aug. 22, 1994, pp. 85–93.

[7] Hil, F. R., *Computer Aided Logical Design With Emphasis on VLSI*, New York: Wiley, 1993.

[8] Preston, J. V., and J. D. Lofgren, "FPGA Macros Simplify State Machine Design," *Electronic Design*, Dec. 5, 1994, pp. 109–118.

[9] Self, K., "Designing With Fuzzy Logic," *IEEE Spectrum*, Nov. 1990, pp. 42–44, 105.

[10] Lewis, M., "Fuzzy Logic: A New Tool for the Control Engineer's Toolbox," *Power Transmission Design*, Sept. 1994, pp. 29–32.

[11] Schwartz, D. G., and G. J. Klir, "Fuzzy Logic Flowers in Japan," *IEEE Spectrum*, July 1992, pp. 32–35.

[12] Sibigtroth, J. M., "Fuzzy Logic for Small Microcontrollers," *Wescon/93 Conference Record*, San Francisco, Sept. 28–30, 1993, pp. 532–535.

[13] Holland, J. M., "Using Fuzzy Logic to Evaluate Environmental Threats," *Sensors*, Sept. 1994, pp. 57–60.

[14] Wright, M., "Neural Networks Tackle Real-World Problems," *EDN*, Nov. 8, 1990, pp. 79–90.

[15] Majors, M., J. Stori, and D.-I. Cho, "Neural Network Control of Automotive Fuel-Injection Systems," *IEEE Control Systems*, June 1994, pp. 31–35.

[16] Johnson, R. C., "Gap Closing Between Fuzzy, Neural Nets," *Electronic Engineering Times*, Apr. 13, 1992, pp. 41, 44.

[17] May, G. S., "Manufacturing ICs the Neural Way," *IEEE Spectrum*, Sept. 1994, pp. 47–51.

[18] Renard, P., "Implementation of Adaptive Control on the Motorola DSP56000/DSP56001," *Motorola APR15/D*, 1992.

[19] De Silva, C., and T.-H. Lee, "Fuzzy Logic in Process Control," *Measurement and Control*, June 1994, pp. 114–124.

[20] Meshkat, S., "Meshkat's Motion—What is an Adaptive Controller?" *Motion*, Mar./Apr. 1988, pp. 20–22.

[21] Stobart, R., "Background to Control," SAE Toptec on Electronic Engine Controls, Dec. 4, 1997, San Francisco, CA.

[22] Khanchustambham, R. G., "A Neural Network Approach to On-Line Monitoring of Machining Processes," http://www.isr.umd.edu:80/TechReports/ISR/1992/MS_92-5/MS_92-5.phtml

[23] Quinell, R. A., "Speech Recognition: No Longer a Dream But Still a Challenge," *EDN*, Jan. 19, 1995, pp. 41–46.

[24] Skinner, T., "Program a Multiprocessor Computer," *Electronic Design*, Jan. 23, 1995, pp. 57–66.

[25] Krohn, N., R. Frank, and C. Smith, "Automotive MPU Architectures: Advances & Discontinuities," *Proc. 1994 Internat'l Congress on Transportation Electronics*, SAE P-283, Oct. 1994, Dearborn, MI, pp. 71–78.

[26] "MC68356—Signal Processing Communication Engine," Motorola Product Brief.

[27] Bursky, D., "No-Compromise Controller Combines DSP, Data Comm," *Electronic Design*, June 13, 1994, pp. 79–88.

[28] AbdelRahman, M., "Artificial Intelligence Expands Sensor Applications," *EC&M*, Dec. 1991, pp. 20–23.

[29] Abdelrahman, M., "Fuzzy Sensors for Fuzzy Logic," *Control Engineering*, Dec. 1990, pp. 50–51.

[30] Maiten, J., et al., "Overview of Emerging Control and Communication Algorithms Suitable for Embedding Into Smart Sensors," *Proc. Sensors Expo*, Cleveland, OH, Sept. 20–22, 1994, pp. 485–500.

8

Transceivers, Transponders, and Telemetry

Tricorder readings indicate an unusually high level of nitrium.
—Lt. Commander Data, from *Star Trek: The Next Generation*

8.1 Introduction

A monumental change that accelerated at the end of the twentieth century and that is also affecting sensing is the desire for portable, wireless products. The proliferation of portable computers combined with wireless communication will have the same effect on sensing that the shift from mainframe to personal computers had on the factory floor. The number of sensed parameters and the ease of obtaining measurements will increase based on remote, wireless sensing. Remote measurements require the combination of the digital technologies (discussed in Chapter 5) and radio frequency (RF) semiconductors, including radio frequency ICs (RFICs). In some instances, RF technology is used to perform the sensing function.

The use of RF technology in remote sensing historically has been associated with geophysical analysis of parameters such as air and surface temperature, wind velocity, and precipitation rate. That information is necessary to predict weather and is typically gathered from equipment mounted on aircraft, satellites, and weather balloons. The applications of RF continue to grow; newer commercial applications that are portable and have much lower cost objectives will be the focus of this chapter.

Portable wireless products require low power consumption to increase their useful life before requiring battery replacement. For measurement of low power consumption, high-impedance sensors, sleep mode, and smart techniques such as the PWM technique (see Chapter 5) are required. Depending on the type of measurement (static or dynamic) and how frequently it must be made, periodic readings can be transmitted to distant recording instruments in process controls, hazardous material monitoring systems, and a variety of mobile data acquisition applications. Such readings would have been more time consuming, dangerous, prohibitively expensive, or difficult with previously available technology.

8.1.1 The RF Spectrum

Recent legislation has expanded the electromagnetic spectrum that is available for wireless communications in the United States. The Federal Communications Commission (FCC) regulation Title 47, part 15, covers unlicensed security systems, keyless entry, remote control, and RF LANs. Table 8.1 identifies key frequency ranges that are already the focus of several industries. The industrial, scientific, and medical (ISM) bands are being used for several applications. Integration is the key to higher performance, smaller packages, and lower cost in the RF arena, as well as previously discussed systems, with several technologies used in various frequency ranges.

Several technology choices exist in the RF front end of a portable communication product, shown in Figure 8.1 [1]. Silicon competes with GaAs in the 1- to 2-GHz range; however, in the 2- to 18-GHz range, GaAs is the only solution. With a supply voltage of 3V, high-frequency (1 GHz) operation of GaAs has an efficiency of 50% versus 40% for silicon bipolar or 43% for a lateral diffused MOS (LDMOS). The tradeoff in efficiency versus cost must be evaluated prior to initiating a custom design. The down converter and low-noise amplifier (LNA)/mixer, transmit mixer, antenna switch, driver and ramp, and power amplifier can be designed with RF chip sets consisting of different IC technologies. The chip set approach can be a cost-effective alternative to higher levels of integration. High-frequency designs for radio frequency are considerably different from high-speed digital processes. The frequency range in RF circuits of 800–2,400 MHz is much higher than the fastest digital circuits. A mixed-signal approach is required that combines not only digital and analog but also RF technology. Circuit isolation is required to prevent unwanted coupling of signals, which can range from 3 V_{p-p} to less than 1 μV_{p-p}. RFICs are being designed for a number of high-frequency applications using a variety of

Table 8.1
RF Applications Versus Frequency

RF Spectrum	Frequency	Miscellaneous
Industrial/scientific/medical	902–928/2,400–2,483.5/5,725– 5,850 MHz	
Microwave radar	10.25, 24, or 34 GHz	
Radar proximity sensor	1 MHz	
SAW in tolls	856 MHz	
Remote key	41 or 230 MHz	
WLAN	2.4 GHz	Data transfer rate (DTR) ≥ 1.6 Mbps
Altair™* WLAN	18 GHz	DTR = 15 Mbps
LAN PCMCIA card	900 MHz	
RF-ID	100 kHz–1.5 MHz and 900 MHz– 2.4 GHz (915 MHz)	
GPS (L1 and L2)	1,575.42, 1,227.60 MHz	
Ultrasonic position sensing	20–200 kHz	
Piston temperature telemetry	1 MHz	
Tire monitoring system	355 MHz or 433.92 MHz	
Medical telemetry (UHF)	450–470 MHz	
Medical telemetry (VHF)	174–216 MHz	
Band	Microwave electronics band frequency	Wavelength
L	1–2 GHz	30.0–15.0 cm
S	2–4 GHz	15.0–7.5 cm
C	4–8 GHz	7.5–3.7 cm
X	8–12.5 GHz	3.7–2.4 cm
Ku	12.5–18 GHz	2.4–1.7 cm
K	18–26.5 GHz	1.7–1.1 cm
Ka	26.5–40 GHz	1.1–0.75 cm
mm	40–100 GHz	0.75–0.3 cm
Ultraviolet		0.1–0.4 μm
Visible		0.4–0.7 μm
Infrared		0.7 μm–0.1 cm

*Altair™ is a trademark of Motorola, Inc.

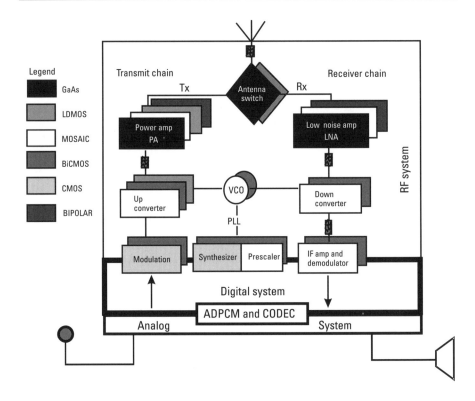

Figure 8.1 RF to digital transition.

technologies, depending on the application. The term *monolithic microwave integrated circuit* (MMIC) is used for RFICs in the microwave range.

The transition from RF to digital baseband is simplified by mixed-signal, analog-digital ICs. As shown in Figure 8.1, a phase-locked loop (PLL) synthesizer and prescaler, as well as a modulator/demodulator (modem), are required for this transition. PLL frequency synthesizers are required in RF applications to provide digital tuning capability for wireless digital communication to implement a cost-effective, multiple-channel design. A PLL allows a precise stable frequency to be generated with an MCU controlling the frequency. Advanced CMOS processes provide the lowest power dissipation and a low operating voltage for 100 to 300 MHz. BiCMOS increases the frequency range (to about 2 GHz) and provides flexible circuit elements at moderate power dissipation. An advanced bipolar process, like the MOSAIC™ (Motorola Self-

Aligned IC) process provides low-power dissipation at a lower cost and has an even higher frequency range than BiCMOS.[1] GaAs provides the lowest operating voltage at the highest cost but can operate above 2 GHz.

Mixed analog and digital process technology is required for the adaptive differential pulse code modulation (ADPCM) coder/decoder (codec) shown in Figure 8.1 that translates analog into a digital format in personal communication devices. The digitized signal is transmitted over the RF channel. The ADPCM is designed for 32 Kbps, which is the codec technology for many personal communication systems worldwide, as well as 64-, 24-, and 16-Kbps data rates. DSP technology is used to achieve optimum data rate and RF bandwidth. Precision analog and high-performance digital technologies are required for the transition from analog to digital regime and back.

GaAs semiconductors are being designed to address the highest speed applications and are well suited to low-voltage operation. However, they are more expensive than silicon and should be used only where their performance would be difficult or impossible to achieve with silicon. The use of GaAs technology adds another dimension to circuit partitioning that must be considered for RF circuits.

8.1.2 Spread Spectrum

Spread spectrum is one of the secure methods of transmitting information via radio signals that recently (1991) has expanded from military use to commercial use. The spread spectrum signal is a much lower magnitude but is generated by a much broader frequency range, 26 MHz compared to a narrowband signal of less than 25 kHz, as shown in Figure 8.2. Narrowband transceivers (combined transmitters and receivers) are highly susceptible to interference by frequency sources near their carrier frequencies. Because they operate within the audio range, the equipment being monitored frequently generates the interference [2].

Spread spectrum offers improved interference immunity, low interference generation, high data rates, and nonlicensed operation at practical power limits [3]. To conform to FCC guidelines, the total transmitted power must be 1W maximum, and the spectral density (the power at any specific frequency) must be no greater than 8 dB in a 3-kHz bandwidth. Two techniques are used to distribute the conventional narrowband signal into a spread spectrum equivalent: direct sequencing and frequency hopping.

---

1. MOSAIC™ is a trademark of Motorola, Inc.

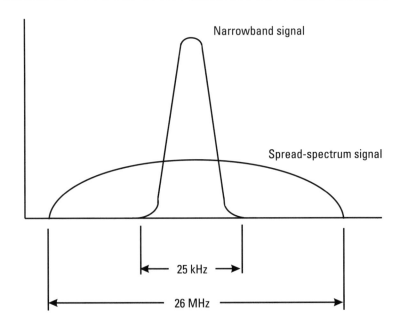

Figure 8.2 Spread spectrum versus narrowband.

Direct sequence is a spread spectrum approach that creates an instantaneous RF signal consisting of many frequencies spread over a portion of the frequency spectrum. The information signal, which is usually digital, is combined with a much faster stream of pseudorandom binary code that is repeated continuously. This technique is difficult to detect, and recapturing the information requires duplicating the spreading code at the receiver end.

Another spread spectrum approach is frequency hopping. Both base and subscriber (handset and base) "hop" from frequency to frequency in a simultaneous fashion [4]. The theory states that noise tends to occur at different frequencies at different times. Therefore, even though a part of the transmission may be lost due to interference, enough of the message will be received by hopping through the interference to create a noticeably better output compared to fixed-frequency systems. See Table 8.1 for a variety of applications and their frequencies. The 1.8- to 2.4-GHz band is open for data collection in Europe, which at present is the second largest market for RF data.

Spread spectrum's ability to put a greater number of carriers in a specific bandwidth and to increase the security of the transmitted data is enabling several new wireless applications for sensing. Spread spectrum techniques combined with new frequency ranges in the unlicensed ISM bands have made new applications not only possible but easy to implement. In addition to communication, commercial applications of spread spectrum include wireless LANs, security systems, instrument monitoring, factory automation, remote bar code reading, automatic vehicle location, pollution monitoring, medical applications, and remote sensing of seismic and atmospheric parameters.

8.2 Wireless Data and Communications

Error-free wireless data is becoming possible due to digital data transmission. A number of protocols are vying for acceptance in RF signal transmission. Existing access methods include time-division multiple access (TDMA), frequency-division multiple access (FDMA), and the newest method, code division multiple access (CDMA). Users are assigned a time slot, frequency, or code that allows multiple users to share the available frequency spectrum. CDMA provides an additional dimension and option that allows more efficient utilization of the available spectrum and therefore greater capacity (up to 20 times greater) with a given bandwidth [5]. That means an increased number of users can access the system and maintain the level of service—access, data rate, bit error rate (BER), and so on—with only a slight degradation in process gain. The discontinuous nature of human speech is the key to one aspect of CDMA technology. Only one-third of the air time is used for speech; the rest is spent on listening and pauses between words, syllables, and thoughts. CDMA uses a variable-rate voice encoder (vocoder), which performs a speech-encoding algorithm to transmit speech at the minimum data rate needed. Encoding rate is from 1 to 8 Kbps, depending on speech activity.

Several techniques, including frequency shift keying (FSK), amplitude shift keying (ASK), on-off keying (OOK), differential phase shift keying (DPSK), quadrature phase shift keying (QPSK), and Gaussian minimum shift keying (GMSK), are used to modulate the signal transmission. Techniques previously used for military and satellite communications such as differential quadrature shift keying (DQSK) and GMSK maximize the information density in a spectral bandwidth. The choice depends on cost and performance objectives within the system.

Cellular digital packet data (CDPD) technology divides data into small packets that can be sent in short bursts during the idle time between cellular voice transmissions. The open specification is designed to support a variety of interoperable equipment and services and is backed by major cellular carriers. CDPD allows faster transmission (19.2 Kbps) of short messages, compared to the existing circuit-switched system, which handles longer files [6]. CDPD also incorporates error correction protocols. Table 8.2 lists some of the competing

Table 8.2
RF Communication: Wireless LAN Protocols and Sponsoring Agencies

Protocol	Sponsor(s)	Comments
X.25 packet protocol	CCITT	Wireless extension of landline specification
802.11	IEEE	Proposed standard; uses ISM frequencies
CDPD	McCaw	Uses unused capacity on cellular voice
RAM Mobitex	Ericcson	Message packet data system
ARDIS™*	Motorola and IBM	Data packets at either 9.6 or 14.4 Kbps

*ARDIS™ is a trademark of Motorola, Inc.

protocols. Several RF-specific terms are defined in the glossary at the end of this book.

8.2.1 Wireless Local Area Networks

Wireless LANs (WLANs) operating at 18 GHz are able to propagate well within a building yet diffuse rapidly outside. FCC licensing is required for WLANs in the digital termination service (DTS) band, 18–19 GHz, which includes the use of low-power radio communications within buildings. Each of the 10-MHz channels within that band can deliver 15 Mbps [7]. Some WLANs also operate in the lower frequency range of 900 MHz to 6 GHz. In addition, there are a variety of infrared links for wireless networks. Those indoor systems have a range of 100m and data rates of 1 Mbps.

8.2.2 FAX/Modems

Real-time sensing using a PCMCIA card fax/modem installed in any portable computer establishes a flexible, easily installed link to a base computer. Up to 28,000 bps are already possible using available technology. PCMCIA cards are also being designed specifically to handle sensor data [8]. One device provides the modem function and a level of intelligence that can obtain information from sensors located in remote or mobile devices.

An error correcting protocol, derived from the asynchronous balance mode of the Consultive Committee of the International Telephone and Telegraph (CCITT)–defined X.25 packet protocol, retransmits only those frames that are not recognized by the receiver (Figure 8.3). The modulation technique easily synchronizes transmitted data. A basic HDLC concept and nonreturn-

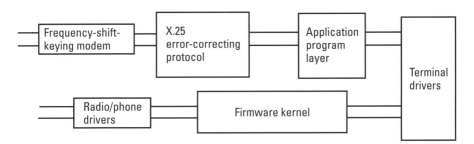

Figure 8.3 Wireless data using a PC card.

to-zero inverted (NRZI) bit patterning increases the clock-recovery data for the receiver. Data is transmitted in packets of up to seven adjustable-length frames. A compression algorithm built into the firmware kernel allows data compression, transfer, and analysis prior to transmission. Error correction capability is achieved by allowing the receiver to store ahead frames that are received out of sequence and to request retransmission of any missing frames.

A file-level protocol facilitates the exchange of data between the mobile unit and the base unit. A microprocessor controls the data transmission and services the serial ports and the parallel port. Global positioning system (GPS) data (discussed in Section 8.3.3), mobile data terminals, and bar code readers/scanners are among the immediate applications for such RF data transmission.

8.2.3 Wireless Zone Sensing

Wireless zone sensors (WZS) have been developed for buildings using the spread spectrum technique. The wireless sensors are part of the modern building automation systems designed for energy efficiency and precise control of many individual areas [9]. Figure 8.4 demonstrates a typical system with wireless zone sensors.

The WZS is a spread spectrum transmitter that provides room temperature and other status information to a receiver located up to 1,000 ft away. A translator in the unit converts the wireless data to a wired communication link. The wired communication link distributes data to variable air volume (VAV) terminal controllers and to the building management system. The VAV adjusts the valve opening to allow more or less flow to the area as required by the WZS. A sensor is matched to its associated VAV by a setup tool connected to each WZS during the installation process. The lithium-battery-powered WZSs are designed to operate an average of two years before the batteries need to be changed.

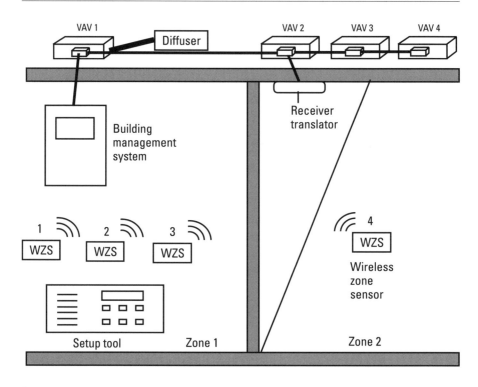

Figure 8.4 Wireless zone sensing.

Wireless data loggers are being used in clean rooms to pinpoint localized humidity fluctuations [10]. In this wireless sensor application, the measurement is required only to locate the source of a problem. Flexibility in choosing the test sites, as well as avoiding clean room suits and drilling holes through the wall, was paramount to design considerations. Again, the low interference possible using spread spectrum technology has provided the wireless transmission of real-time data in an environment that previously prohibited radio transmission. Up to 20 data loggers have been linked to a single wireless base station. As a result, a single PC can handle the measurements from a number of widely separated data collecting units.

8.2.4 Optical Signal Transmission

The infrared spectrum is also used for transmitting data. It is low cost for cost-sensitive applications and does not require licensing. However, it is limited to line-of-sight (LOS) transmission and cannot penetrate solid objects. The infrared spectrum (indicated in Table 8.1) is from 0.7m to 0.1 cm.

8.3 RF Sensing

RF techniques are also used in wireless sensors. RF sensors are a noncontact and intrinsically safe way to measure velocity and distance, detect motion and pressure, and indicate direction of motion. In addition, RF sensors are used for liquid-level sensing and detecting the presence of foreign objects. RF sensing techniques include SAW sensors, Doppler radar, sonar, ultrasonics, and microwave sensors. Some of these sensing techniques are used in vehicle antitheft and remote entry systems.

8.3.1 Surface Acoustical Wave Devices

SAW technology is being used for both RF communication and sensing. SAW devices are used to enhance the stability of oscillators in transmitters and as front-end filters to set the bandwidth and improve intermodulation performance of receivers. SAW devices respond to temperature, pressure, force, and vibration. The SAW device placed in an oscillator circuit, as shown in Figure 8.5, provides a frequency variation that can be measured accurately [11]. Differential techniques can be used to compensate for unwanted stimuli. The frequency range can be from direct current to several gigahertz.

SAW devices are also being used in chemical sensors. In gas sensors, dual SAW oscillators are frequently used. One SAW device acts as the reference, and the other has a gas-sensitive film deposited between the input and output interdigitated transducers. The relative change in the SAW oscillator frequency is directly proportional to gas concentration. In one study, porous oxide coatings were used as the discriminating element in the sensor [12]. The coating microstructure was evaluated by monitoring nitrogen adsorption-desorption at 77K using SAW devices as sensitive microbalances. The frequency shift of a SAW device coated with zeolite in a sol-gel microcomposite film was –6,350 Hz for methnol, –10,200 Hz for isopropanol, and +74 Hz for iso-octane.

Automotive navigation systems are also potential users of SAW devices. The SAW device is used in an RD-identification (RD-ID) tag to identify a vehicle for billing purposes. In that application, SAW devices operate at 856 MHz (discussed in Section 8.3.7).

8.3.2 Radar

Radar (the acronym for *radio detecting and ranging*) uses reflected radio waves to measure range, bearing, and other parameters. A recently developed radar proximity sensor (Figure 8.6) uses spread spectrum techniques and is a good example of the critical elements of a radar system [13]. This particular sensor is

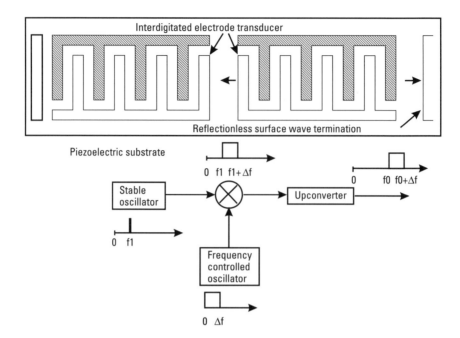

Figure 8.5 SAW device.

designed to meet cost-sensitive applications such as blind-spot detection in automobiles. The approach detects the echoes that reflect back over a limited range from continually propagating electromagnetic impulses. The sensor is not affected by most materials, so its location is not a factor in its ability to function properly. The unit can see through walls and distinguish between wood and metal wall. The radar noise is coded so an unlimited number of sensors can be colocated without interfering with each other.

The proximity sensor's antenna transmits a subnanosecond pulse at a noise-dithered repetition rate. The pulse can be from 50 ps to 50 ns long. Dithering randomizes the time of the transmission in a spread spectrum that appears as noise to other detectors. The echoes returned to the (patented) receiver are sampled typically at a frequency of 1 MHz and in a time frame (plus delay) determined by the transmitter. The range of the sensor is limited to 20 ft (6m) or less. However, several applications are possible with the sensor, including car security, voice-activated navigation, and detectors for locating one material embedded in another, such as steel within concrete.

Production microwave proximity sensors operate on radar principles. A simple microwave sensor consists of a transceiver, an antenna, and signal

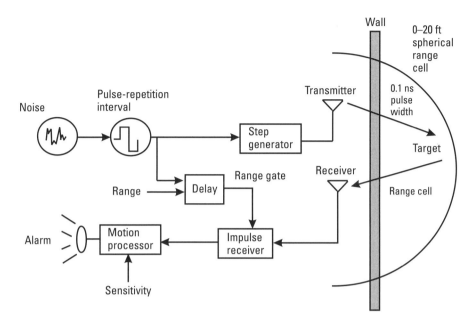

Figure 8.6 Radar sensor schematic.

processing circuitry [14]. See Table 8.1 for the microwave frequency range and its various subclassifications.

The transceiver in a microwave sensor uses a Gunn diode or a field effect transistor (FET) to generate a low-power microwave signal. The antenna focuses the energy developed into a beam, the size of which is determined by the application. The reflection of the beam from an object provides a lower level signal, which can be analyzed by the signal processing circuitry. The signal can be used for Doppler shift (motion, speed, direction), strength (presence), or phase change (distance), depending on the application.

Microwave sensors are experiencing size reductions and performance improvements from the integration possible in MMIC chips. An MMIC can have all of the circuit or large portions of it on a single chip. Flip-chip packaging (see Chapter 10) of an MMIC containing the detector diodes has allowed frequency operation up to 40 GHz [15].

8.3.3 Global Positioning System

GPS is based on information supplied by 24 satellites located in six orbital planes [16]. The satellites pass over the earth at an altitude of 20,183 km

(10,898 nautical miles). A single satellite orbits the earth twice for each earth rotation, tracing exactly the same path twice each day but passing four minutes earlier than it did the day before. The design of the system ensures that at least four satellites are in view at any one time, on or above the surface of the earth under all weather conditions.

GPS satellites transmit at two frequencies: L1, centered at 1,575.42 MHz, and L2, centered at 1,227.6 MHz. Each satellite broadcasts a navigation message that includes a description of the satellite's position as a function of time, an almanac, and clock correction terms. Each message comprises 25 frames, each 30 seconds long. Commercial GPSs are capable of locating to within 100m.

GPS technology is an integral part of the automotive ITS (discussed in Section 8.3.6). GPS can be used to track field personnel and for locating position relative to location on a CD-ROM map. GPS products have been developed for consumer use. In addition to the RF components necessary to convert the signals from satellites, a 32-bit microprocessor is also required to compute the position algorithm quickly and provide frequent updates to the user. The complex sensor (receiver) is contained within a board that is 50.8 mm by 82.6 mm by 16.3 mm [17]. Increased acceptance and application in high-volume systems such as the ITS will allow further component reduction through increased integration.

8.3.4 Remote Emissions Sensing

The remote measurement of vehicle exhaust emissions is one way that remote sensing will start to affect a broad range of the population. Identifying vehicles that emit high levels of carbon monoxide and hydrocarbon is possibly the next step toward reduced emissions from vehicles with combustion engines. Figure 8.7 illustrates a remote sensing system [18]. The main parts of the system include an infrared (IR) detector and source; a video camera to record the license plate number; a modified police radar gun; and a personal computer with specially developed software.

The remote sensing device (RSD) system operates by continuously monitoring the intensity of an IR source. The presence of a vehicle is indicated when the beam is broken, resulting in the reference voltage dropping to zero and a span voltage measurement being made. The value prior to beam interruption is also stored. As the vehicle exits the beam, samples are taken at 125 Hz for over 1 second. The carbon dioxide spectral region is filtered at 4.3 μm and the carbon monoxide region at 4.6 μm to isolate those values. Hydrocarbon values can be detected as well. The system incorporates three noncontact remote

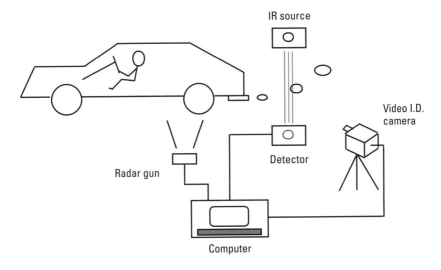

Figure 8.7 Remote measurement of vehicle exhaust emissions.

measurement devices, including an extension of previously used police radar technology, which typically has a maximum range of 800m.

8.3.5 Remote Keyless Entry

Available wireless remote control of door locks on vehicles is a step toward an automatic driver sensor (ADS). The ADS will be able to identify that a specific driver is approaching the vehicle and that the driver will require access and possibly control settings different from those used by the previous driver. Today's remote entry systems have either IR or RF sensors. One RF system, diagrammed in Figure 8.8, has a transmitter in the ignition key that is powered by a lithium battery [19]. The transmitter generates a frequency of about 41 or 230 MHz when a switch in the key is depressed. The vehicle's rear window defogger acts as the antenna. The antenna receives the signal, which is then sent to the receiver for amplification, FM detection, and wave modification. If the transmitted code matches the stored code, the motor drive circuit either unlocks or locks the doors. The wireless system can be activated within 3 ft of the vehicle. It has a standby mode, which is intermittent to avoid excessive drain on the vehicle's battery. If 10 or more incorrect codes are transmitted to the receiver within a 10-minute period, the system reacts as if a theft is occurring and all reception is discontinued. The key must be inserted manually to verify ownership.

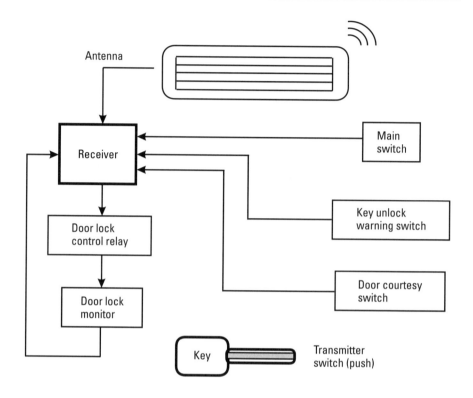

Antenna

Receiver

Main switch

Key unlock warning switch

Door lock control relay

Door courtesy switch

Door lock monitor

Key

Transmitter switch (push)

Figure 8.8 Remote door lock control system.

8.3.6 Intelligent Transportation System

The ITS, formerly known as intelligent vehicle highway system (IVHS) in the United States and by other names (and nicknames, including "smart cars" and "smart highways") around the world, requires several sensing technologies, many of which are RF based, to implement full functionality. ITS has defined several areas that are being addressed to varying degrees, from research to production systems. These areas attempt to solve a number of system-related problems as well as provide new functions and services to drivers. The application areas include advanced traffic management system (ATMS), advanced traveler information system (ATIS), advanced public transportation system (APTS), advanced vehicle control system (AVCS), advanced rural transportation system (ARTS), and commercial vehicle operation (CVO).

The various systems require a number of sensors, including GPS sensing, closed-circuit TV monitoring, IR detectors, dead reckoning, automatic vehicle location, and identification [20]. Dead reckoning can be accomplished by differential wheel-speed sensing and a fluxgate magnetic compass or a combination of inclinometers, gyroscopes, inverse Loran, electronic odometers, and/or antilock brake system wheel-speed sensors. In addition, the cellular infrastructure and/or radio beacons play an integral part in the vehicle navigation requirements of ATMS, ATIS, APTS, ARTS, and CVO. The AVCS system will require sensors to determine distance between vehicles for smart cruise control (automatic vehicle spacing) and closing distance (time to impact) for collision avoidance systems. Figure 8.9 indicates the typical frequency ranges of the RF sensors.

One configuration of a vehicle navigation and warning (VNAW) system is shown in Figure 8.10. The smart inertial navigation system (SNS) contains the computing that provides filter and integration for the GPS input as well as the accelerometer and gyroscope inputs [21]. The output of the SNS is position information for the VNAW system.

Collision avoidance in an automotive ITS uses a 75-GHz three-beam radar unit to provide a time-to-impact warning. One system can track up to 12 objects while measuring range angle and relative velocity [22]. The radar system also employs a video camera mounted near the inside mirror to provide a 30-degree field of view. The camera detects lane markings, road edges, and objects in the vehicle's path.

Near-obstacle detection systems (NODSs), for parking and blind-spot detection, use Doppler radar operating at 10.5 to 24 GHz to sense objects

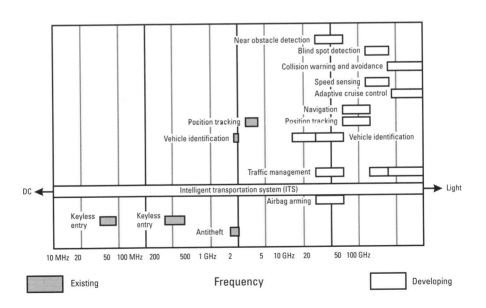

Figure 8.9 Frequencies of RF automotive sensing applications.

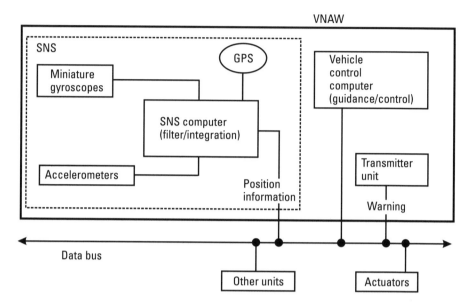

Figure 8.10 VNAW system.

within a few inches or feet of a vehicle. Sonar has also been developed for the shorter range measurements. Sonar units are mounted at each corner of the vehicle. One sonar device operates at 50 kHz and drives a 300V transducer to generate the sonic pulse.

True road speed is one of the inputs that is required in the vehicle navigation portion of ITS. A microwave Doppler radar sensor mounted at the front of the vehicle can provide that measurement. Intrinsic errors due to the spread of the Doppler frequencies from the top to the bottom of the beam, sensor location, false signals from clutter, pitching, vibration, signal glint (irregular ground reflections), short-term oscillator instability, and dirt are among the error possibilities for this speed-sensing approach [23]. Microwave sensors are not significantly affected by humidity, temperature, and air movement and do not require isolation of transmitter and receiver, which are considerations for ultrasonic Doppler sensing.

The complete ITS contains a number of other systems that ultimately accomplish a number of goals, including safer travel on urban and rural roads and higher usage of the existing highway system. Certain systems and subsystems can and are being implemented separately. One element of ITS already achieving high volume and usage in a variety of other applications is RF-ID tags.

8.3.7 RF-ID

RF-ID tags are being used to track inventory in warehouses; work in process (WIP) in manufacturing plants; and animals in laboratories, on farms, or in the wild; as well as for automatic toll collection for vehicles. The RF tag is a transponder that is read and decoded by an RF reader. For inventory and WIP in harsh environments or non-LOS operations, RF-ID is an alternative to a barcode system [24]. The tags can be active or passive. Active tags have onboard batteries and larger data capacities approaching 1 Mb. Decision making ability is included in some varieties with the addition of a microcontroller. Passive tags obtain their operating power from RF energy transmitted from an antenna and as a result have limited data capacities, typically 1 to 128 bits.

RF-ID systems operate in both low- and high-frequency ranges. Low-frequency systems typically operate in the frequency range of 100 kHz to 1.5 MHz and also have lower data transfer rates. They are cost effective in access control and asset tracking applications. High-frequency systems operate in the spread spectrum range of 900 MHz to 2.4 GHz and have higher data transfer rates. High-frequency systems cost more than low-frequency systems, but they can operate at distances up to 100 ft (30m).

The main blocks of a passive RF-ID tag are shown in Figure 8.11 [25]. An RF burst is received, rectified, and used to charge a capacitor that serves as the power supply for the tag circuitry. A voltage regulator keeps the voltage across the capacitor at a constant 2V. Logic circuitry interprets the command from the RF burst. An interrogation request is answered with transmission of a number from the EEPROM. Information in the EEPROM can be changed by

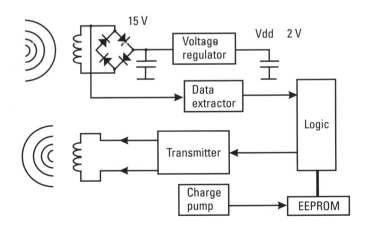

Figure 8.11 RF ID block diagram.

a transmission that enables the logic to store new data into the onboard memory. The charge pump increases the 2V to 15V to reprogram the EEPROM.

RF-ID tags are being used for automatic toll collection. The driver does not have to slow down for the system to detect the vehicle's ID and debit the driver/owner's account. Congestion on highways is reduced and revenue still is collected to cover road construction and maintenance.

8.3.8   Other Remote Sensing

RF bar code readers, which have become commonplace, demonstrate the freedom of operating untethered for a commercial application. Available RF units provide data from a charge-coupled device (CCD), laser scanner, or infrared wand sensor to a base station that can be up to 150 ft (45.7m) away. The units are mobile and can be used without the interference of wires in high-traffic areas.

Remote meter reading is being accomplished by transmitting the amount of gas, water, or electricity consumed at a particular site to a mobile receiving unit. A single operator can verify the status of almost twice the number of accounts, thereby reducing the cost of the measurement. Remote bar code readers and scanners are also among the applications that are taking advantage of RF technology in a sensing environment.

8.3.9   Measuring RF Signal Strength

Exposure, especially continued exposure, is a concern for RF signals. The American National Standards Institute (ANSI) has limits for equivalent permissible exposure, depending on the frequency. An RF dosimeter has been designed to detect and record the strength and duration of electric fields present in work areas of naval vessels [26]. The potential to include the electronics in the already pocket-sized sensor with other RF measurement techniques may provide useful data in future applications.

8.4   Telemetry

Telemetry is a remote measurement technique that permits data to be interpreted at a distance from the primary detector. Telemetry is used in race cars to allow the pit crew to analyze the real-time data generated from a vehicle on the track and provide feedback to the driver that can affect the outcome of the race. For example, in Ford's Formula One system, more than a dozen sensors placed around the vehicle provide information to the engine control computer and transmit the data to a mobile laboratory that travels to the races [27]. An indication of a potential problem can initiate a pit stop before the problem becomes a disqualifying failure. Indianapolis 500 and National Association of Stock Car Auto Racing (NASCAR) vehicles use similar telemetry. In fact, proving-ground vehicles equipped with telemetry systems allow automotive engineers to evaluate, from their offices, development vehicles on high-speed test tracks. Cellular communications with drivers provide on-the-spot direction for performing various tests with real-time feedback on how systems are affected. In both racing and vehicle development, time is short and cost investment high. Telemetry makes the outcome more predictable.

Telemetry is also used on production vehicles in systems that monitor pressure and temperature for each tire. The RF transmitter at each wheel sends a signal (355 or 433.92 MHz are common frequencies) that is received by a unit mounted under the dash, and a dashboard-mounted display provides information to the driver. One system uses pulse code modulation to transmit digital data by turning the carrier frequency on or off, producing a burst of radio frequency energy [28]. Switching rate and time are controlled to create the code.

Wheel-mounted transmitters deliver 10 frames of coded data in 128-ms bursts at approximately 30- to 35-sec intervals. Each frame has 8 bits of data and a blanking period. Two bits indicate tire pressure, two identify the wheel being measured, and the other four identify the car model and year. Tire pressure and temperature are important vehicle measurements that not only affect the performance and economy of the vehicle but also have a significant impact on vehicle safety. These systems may be common on future vehicles due to the combined capability and continued cost reduction of integrated sensing and RF technology.

Another difficult-to-get-at measurement is the temperature of the piston in an internal combustion engine. Telemetry has been used to transmit the temperature indicated by the variation of a temperature-sensitive chip capacitor. The system schematic is shown in Figure 8.12 [29]. Increasing temperature decreases the capacitance, which in turn increases the transmitting frequency. A multiloop antenna in the oil pan receives the transmitted signal. Data is converted to temperature through a calibration curve. The system has seven data channels to map temperatures at a number of locations inside the piston. A small 22-pin hybrid package contains the multiplexer and the blocking oscillator. Power is generated based on the piston's movement. A minimum engine speed of 1,200 rpm is necessary, and temperatures from 150 to over 600°F (65–315°C) can be measured. To keep track of the data being sent, a reference capacitor that is not temperature dependent transmits a 1-MHz signal. The remaining channels are each turned on for 1 sec and then off for 1 sec.

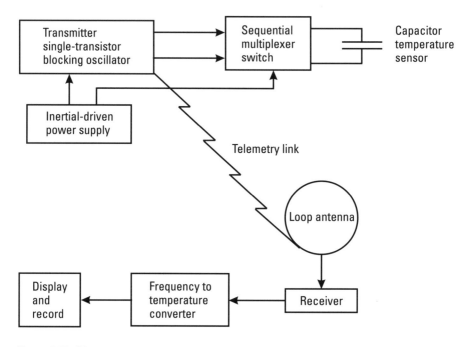

Figure 8.12 Piston temperature telemetry system.

Temperature-dependent frequencies are generated between 250 kHz and 500 kHz for the data channels.

Ambulatory monitoring of critically ill patients or patients requiring real-time diagnostics for analysis purposes has increased the use of medical telemetry. This telemetry can be short range within a hospital floor, or long range for an entire wing. Most telemetry is within the 174–216 MHz (VHF) frequency band for ultra-low power transmission. Higher power requirements use an alternative UHF band (450–470 MHz) and require licensing.

Real-time monitoring of noninvasive blood pressure, partial pressure of oxygen, and peripheral pulse provide additional diagnostic and assessment information on many cardiac and respiratory patients. Electrocardiograms (ECGs) and the partial pressure of oxygen can be monitored by two cigarette-pack-sized units weighing a total of only 380g [30]. A central station receives the data, analyzes the data for anomalies, and routes waveform data and parameter data to additional analysis or recording equipment. Consolidating several patients into one remote station can lower hospital costs by reducing the staff for intensive care and postoperative units, as well as provide an improved level of patient care. More data can be accumulated and analyzed with less effort from nurses and doctors. In addition, the onset of problems can be detected quickly and appropriate action taken sooner.

RF telemetry has also been investigated for microminiature transducers for biomedical applications. Transferring data and power into and out of the body to implanted transducers is a critical area because of the reliability of the component and, more important, the restrictions and potential of infection to the patient. A microstimulator has been developed that measures only 1.8 mm^3 by 1.8 mm^3 by 9 mm^3 using RF telemetry operating at 1 MHz for power and control [31]. The assembly includes a micromachined silicon substrate that has the stimulating electrodes, CMOS, and bipolar power regulation circuitry; a custom-made glass capsule electrostatically bonded onto the silicon carrier to provide a hermetically sealed package; hybrid chip capacitors; and the receiving antenna coil. The application of microtelemetry will be more practical with improvements in integrated micromachining and reduction in the receiving antenna size.

8.5 RF MEMS

MEMS technology is being used to develop RF components that demonstrate superior high-frequency performance relative to conventional (usually semiconductor) devices. These new devices provide potential for new system capabilities [32]. Darpa has investigated MEMS technology for radio front ends, capacitor banks, and time-delay networks for quasi-optical beam steering and reconfigurable antennas.

A recently issued patent describes a monolithically integrated switched capacitor bank using MEMS technology that is capable of handling gigahertz-signal frequencies in both the RF and millimeter bands [33]. In addition, the technique maintains precise digital selection of capacitor levels over a wide tuning range. Each MEMS switch includes a cantilever arm suspended above the substrate that extends over a ground line and a gapped signal line with a set of contacts on the arm and the substrate. The MEMS switch is actuated by a voltage applied to the top electrode that produces an electrostatic force and attracts the control capacitor structure toward the ground line, causing the electrical contact to close the gap. The integrated MEMS switch-capacitor pairs have a large range between their on-state and off-state impedance and exhibit superior isolation and insertion loss characteristics.

Antennas, transmission lines, and other RF components are being built using micromachining techniques [34]. The new micromachined chips are less expensive than previous silicon versions. Figure 8.13 shows a 40-GHz

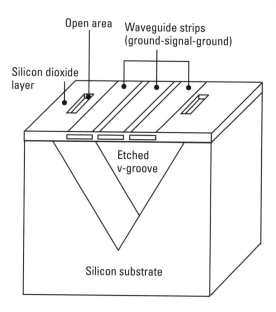

Figure 8.13 40-GHz micromachined silicon waveguide. (*After:* [34].)

waveguide [34]. Standard CMOS processing was used with a postprocess etch to obtain the V-groove and open area. The transverse electromagnetic mode (TEM) waveguide provides low-loss and high-phase velocity at up to 40 GHz.

The cost of installing sensors, as much as $2,000/ft in some situations (such as portions of an aircraft), is instigating research into wireless and wireless MEMS sensors [35]. Researchers at the University of California, Los Angeles, have been investigating low-power wireless integrated microsensors (LWIMs) for distributed-microsensor networks [36]. They have demonstrated high-selectivity micropower receiver systems using CMOS SAW-stabilized receivers and transmitters, a monolithic accelerometer and suspended inductor antenna loop, and a wireless infrared sensor. Ongoing efforts to combine those systems and subsystems through packaging or monolithic IC technology should provide new options for future distributed networks.

8.6 Summary

Real-time data acquisition systems will increasingly look to RF communications for faster installation and easier maintenance. A variety of wireless data services are being developed that savvy systems developers can use to create new sensing systems. Field service workers and other mobile data collectors who need to communicate the results of sensed parameters or acquired data will not want to be connected by wires to a distributed system. Home, office, and industrial monitoring systems are easier to implement using RF signal transmission than using hardwired installations, even with simple two-wire systems. Also, several hazardous monitoring situations require sensing and transmitting of data to minimize exposure of humans to toxic material. The combination of wireless technology with MEMS technology is part of an exciting new area for smart sensors.

References

[1] Frank, R., "Improved Portable Communications Through Low Voltage Silicon Design," *Portable by Design Conf.*, Santa Clara, CA, Feb. 14–18, 1994, pp. PC23–PC34.

[2] Vilbrandt, P., "Wireless Data Communications," *Sensors*, May 1993, pp. 19–21.

[3] Gaston, D., "Spread Spectrum Systems: Evaluating Performance Criteria for Your Application," *Proc. Second Annual Wireless Symposium*, Santa Clara, CA, Feb. 15–18, 1994, pp. 489–507.

[4] 1994 Motorola Communications Resource Guide, BR1444/D, Motorola, Inc.

[5] Leonard, M., "Digital Domain Invades Cellular Communications," *Electronic Design*, Sept. 17, 1992, pp. 40–52.

[6] Phillips, B., "Pumping Data Into Cellular," *OEM Mag.*, Sept. 1994, pp. 32–41.

[7] Mathews, D. J., and C. L. Fullerton, "Microwave Local Area Network for the Computer Office," *Applied Microwave*, Winter 91/92, pp. 40–50.

[8] Nass, R., "Error-Free Wireless Data Transmission Can Be Embedded in a PCMCIA Card," *Electronic Design*, Dec. 5, 1994, p. 48.

[9] Alexander, J., R. Aldridge, and D. O'Sullivan, "Wireless Zone Sensors," *Heating/Piping/Air Conditioning*, May 1993, pp. 37–39.

[10] "Wireless Data Loggers Help Pinpoint Cleanroom Humidity Fluctuations," *Microcontamination*, Aug. 1994, p. 36.

[11] Elachi, C., *Spaceborne Radar Remote Sensing: Applications and Techniques*, New York: IEEE Press, 1988.

[12] Frye, G. C., et al., "Controlled Microstructure Oxide Coatings for Chemical Sensors," *IEEE Solid-State Sensor and Actuator Workshop*, Hilton Head, SC, June 4–7, 1990, pp. 61–64.

[13] Ajluni, C., "Low-Cost Wideband Spread-Spectrum Device Promises to Revolutionize Radar Proximity Sensors," *Electronic Design*, July 25, 1994, pp. 35–38.

[14] "Microwave Sensors," *Measurements & Control*, Dec. 1992, p. 173.

[15] Fischer, M. C., M. J. Schoessow, and P. Tong, "GaAs Technology in Sensor and Baseband Design," *Hewlett-Packard Journal*, Apr. 1992, pp. 90–94.

[16] Harris, C., and R. Sikorski, "GPS Technology and Opportunities," *Expo COMM China '92*, Beijing, Oct. 30–Nov. 4, 1992.

[17] VP Oncore GPS Receiver, Motorola Brochure, 1994.

[18] Glover, E. L., and W. B. Clemmens, "Identifying Excess Emitters With Remote Sensing Device: A Preliminary Analysis," SAE 911672, Warrendale, PA.

[19] McCarty, L. H., "Coded Radio Signal Locks/Unlocks Car's Doors," *Design News*, Jan. 22, 1990, pp. 108–109.

[20] Sweeney, L. E., Jr., "An Overview of IVHS Sensor Requirements," *Proc. Sensors Expo West*, San Jose, CA, Mar. 2–4, 1993, pp. 229–233.

[21] Maseeh, F., "Microsensor-Based Navigation and Warning Systems: Applications in IVHS," *Proc. Sensors Expo West*, San Jose, CA, Mar. 2–4, 1993, pp. 251–255.

[22] Sawyer, C. A., "Collision Avoidance," *Automotive Industries*, Jan. 1993, p. 53.

[23] Kidd, S., et al., "Speed Over Ground Measurement," SAE Technical Paper 910272, Warrendale, PA.

[24] Rishi, G., "RF Tags in Manufacturing," *ID Systems*, Nov. 1994, pp. 51–54.

[25] McLeod, J., "RF-ID: A New Market Poised for Explosive Growth," *Electronics*, Feb. 8, 1993, p. 4.

[26] Rochelle, R. W., et al., "A Personal Radio-Frequency Dosimeter With Cumulative-Dose Recording Capabilities," *Proc. Sensors Expo 1990*, Chicago, Sept. 11–13, 1990, pp. 107B-2-9.

[27] "Telemetry: Racing Into Your Future," Ford Electronics Brochure, Dearborn, MI.

[28] Siuru, W. D., Jr., "Sensing Tire Pressure on the Move," *Sensors*, July 1990, pp. 16–19.

[29] Murray, C. J., "Telemetry System Monitors Piston Temperatures," *Design News*, 10-2-89, pp. 192–193.

[30] "Medical Telemetry to Wireless," *Medical Electronics*, Oct. 1993, pp. 106–107.

[31] Akin T., et al., "RF Telemetry Powering and Control of Hermetically Sealed Integrated Sensors and Actuators," *IEEE Solid-State Sensor and Actuator Workshop*, Hilton Head, SC, June 4–7, 1990, pp. 145–148.

[32] Brown, E. R., "RF-MEMS Switches for Reconfigurable Integrated Circuits," *IEEE Trans. Microwave Theory Tech.* (1998), 46(11, Pt. 2), pp. 1868–1880.

[33] "Monolithically Integrated Switched Capacitor Bank Using Microelectromechanical System (MEMS) Technology," US Patent No. 5,880,921, issued Mar. 9, 1999.

[34] Robinson, G., "Process Builds Silicon RF Chips," *Electronic Engineering Times*, Nov. 25, 1996, pp. 37, 40.

[35] Mangers, W. M., G. O. Allgood, and S. F. Smith, "It's Time for Sensors to Go Wireless—Part 1: Technology Underpinnings," *Sensors*, Apr. 1999, pp. 10–20.

[36] Bult, K., et al., "Wireless Integrated Microsensors," *Proc. Sensors Expo*, Boston, 1996, pp. 33–38.

9

MEMS Beyond Sensors

"And these other instruments, the use of which I cannot guess?"
"Here, Professor, I ought to give you some explanations. Will you be kind enough to listen to me?"
                                        —Jules Verne, *Twenty Thousand Leagues Under the Sea*

9.1 Introduction

The micromachining technology that has enabled semiconductor sensors is being applied to control systems for numerous mechanical applications. In some cases, the microstructures that are being developed have no direct relationship to sensors. However, many of those devices enhance the performance of the total system or allow system design that was not previously possible. Because micromachining is fundamental to manufacturing those structures, future developments in that area will improve both micromachining technology and the systems that utilize them.

The microscale design and fabrication of mechanical and electrical structures is called *microelectromechanical systems* (MEMS). The terms *microsystems* and *microsystems technology* (MST) are also used to describe the structures formed using micromachining, especially in Europe. The term *intelligent MEMS* (IMEMS) describes a monolithic MEMS and IC combination. The extent to which micromachining is utilized to produce components that are not sensors is demonstrated by the list of developed silicon structures shown in Table 9.1 [1]. The batch processing of mechanical components has the same potential for mechanical engineering (and other disciplines) that

201

Table 9.1
Micromechanical Structures in Silicon (*After:* [1])

Cryogenic microconnectors	Micromotors
Fiber-optic couplers	Micropneumatic tables
Film stress measurement	Micropositioners
Fluidic components	Microprobes
IC heatsinks	Micropumps
Ink jet nozzles	Microreactor
Laser beam deflectors	Microrelays
Laser resonators	Microrobotic arm
Light modulators	Microrobots
Magnetographic print heads	Micro-SEM
Membranes	Microscanners
MEMS PC card	Microspectroscopy
Microaligners	Microspeakers
Microbalances	Microswitches
Microbolometer (IR detector)	Microstimulator
Microchannels	Microtips
Microchromatograph	Microtransmissions
Microengines	Microvacuum tubes
Micro-Fresnel lens	Microvalves
Microfuses	Microvalve arrays
Microgears	Microwaveguide
Microgripper/microtweezers	Nerve regenerators
Microinterconnects	Photolithography masks
Microinterferometer	Pressure switches
Microlocks	Pressure regulators
Micromachined laser diodes	rms converters
Micromanipulator/microhandler	Thermal print heads
Micromechanical memory	Thermopile
Micromicrophone	Torsion mirrors
Micromirrors	Vibrating microstructures
Micromolds	Wind tunnel skin friction

semiconductor batch processing has had for electrical engineering. Today's $150-billion-plus semiconductor industry and resulting electronics industry would not exist without batch processing technology. This chapter discusses a few of the areas that are being explored to indicate the variety and the extent of MEMS technology. The MEMS field is progressing and expanding very fast. The examples given show the broad coverage of this new field. New actuator designs and new MEMS technologies are being developed rapidly. Readers interested in the most recent developments are referred to the journals, books, and Web sites referenced in this chapter and at the end of the book.

9.2 Micromachined Actuators

Actuators micromachined from silicon and other semiconductor materials use electrostatic, electrothermal, thermopneumatic, electromagnetic, electroosmotic, electrohydrodynamic, shape memory alloy (SMA), thermoresponsive polymers, and other means to provide motion. Table 9.2 is a comparison of the capabilities of several actuators [2], ranging from existing products that provide unmatched performance compared to their macroscale counterparts to intriguing lab curiosities that require significant development to become practical. Examples of a silicon microvalve, micromotors, micropump, microdynamometer, microtransmission, and a microsteam engine, as well as actuators in alternative materials, are discussed in this section.

9.2.1 Microvalves

One area of MEMS actuators that has achieved production status is microvalves. Figure 9.1 shows one design, a silicon Fluistor™ (or fluidic transistor) microvalve that is approximately 5.5 mm by 6.5 mm by 2 mm [3].[1] The bulk micromachined cavity in the top section is filled with a control liquid. In the unactivated state, gases can flow through the valve. A voltage applied to the heating element on the diaphragm causes sufficient expansion of the liquid to deflect the diaphragm, close the valve seat, and restrict flow. The valve has a dynamic range of 100,000 to 1, controlling gas flows from 4 μL per minute to 4L per minute at a pressure of 20 psi.

9.2.2 Micromotors

Micromotors are among the more interesting demonstrations of the future potential of MEMS. Several researchers have fabricated electrostatic and

---

1. Fluistor™ is a trademark of Redwood Microsystems, Inc.

Table 9.2
Summary of Actuator Driver Requirements (*After:* [2])

Concept	Actuation Strength	Actuation Time	Input Power Requirements
Thermopneumatic	34 kPa	0.03 sec	2.5W
Thermoresponsive polymer	437 kPa	0.05 sec	30 mW
Phase change	100 kPa	0.04 sec	1.9 mW
Thermal blocking	100 kPa	0.015 sec	3W
SMA	150 kPa	0.2 sec	0.12A
Bimetallic strip	50 kPa	1.0 sec	0.5A
Dielectric heating	4 Pa	0.02 sec	10V @ 4 MHz
Capacitive	50 kPa	md*	2700V
Piezoelectric	25 kPa	md	1000V
Electrohydrodynamic	2.5 kPa	0.0004 sec	700V
Interfacial tension	10 kPa	0.002 sec	1V
Magnetorestrictive	50 kPa	md	72A
Two coils	50 kPa	md	18A
Ferromagnetic film	50 kPa	md	1.4A
Permanent magnet	300 kPa	md	0.3A

*md = membrane dominated

electromagnetic motors. An example of an electrostatic motor is shown in Figure 9.2 [4]. The rotor's diameter is typically about 0.1 mm. The stator is activated by pulses that produce an electrostatic force. Coupling the motor to a load and friction are among the problems that must be solved to make the motor useful. However, a fan for cooling ultrahigh-performance microprocessors is among the possibilities that can be envisioned for this technology. *Tribology* is the study of wear and is among the areas receiving considerable investigation for these machines, which cannot be lubricated using conventional approaches and which maintain the unique capabilities that MEMS technology promises.

The coil windings in a magnetic micromotor require a thicker actuator cross-section. One approach uses a polyimide-based process that allows microstructures to be fabricated on top of a standard CMOS process [5]. The core of the motor is multilevel electroplated nickel-iron wrapped around a meander

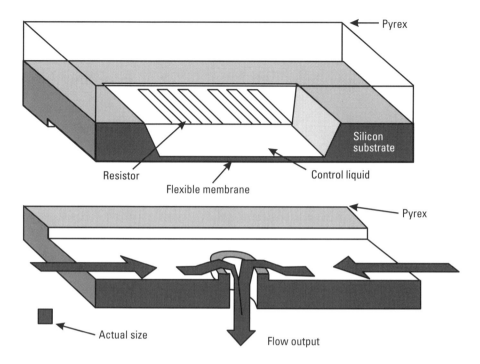

Figure 9.1 Microvalve construction and flow path. (Courtesy of Redwood MicroSystems, Inc.)

conductor. The meander structure for the conductor and the coil is accomplished by reversing the normal roles of the conductor and magnetic core. The magnetic core is wrapped around a planar conductor either by interlacing or by interconnecting multilevel metal layers. Polyimide is used as the dielectric interlayer between for embedding the coils and the cores. Figure 9.3 shows the functional micromotor and an example of the meander coil [5, 6]. The rotor was operated up to a speed of 500 rpm based on the limitations of the drive controller.

Surface micromachining technology has been used to fabricate a micromachined wedge stepper motor [7]. As shown in Figure 9.4, the outer ring has internal teeth, which are engaged by a tooth on each end of the reciprocating shuttle. The outer ring can have gear teeth on the outside or other means of engaging another mechanical member. The figure shows a rotary version of the three-poly layer indexing stepper motor. However, a linear version is possible as well. This approach can be used in designs that require precise and repeatable positioning of micromechanical components.

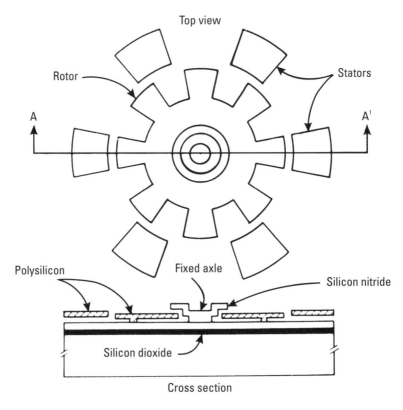

Figure 9.2 Electrostatic micromotor. (*After:* [4].)

9.2.3 Micropumps

A miniature peristaltic pump has been designed, fabricated, and tested. The pump consists of three silicon wafers bonded together, as shown in Figure 9.5, to produce a flow channel, membrane, and heater [8]. Silicon fusion bonding was used to bond the wafers containing those elements. The heaters are suspended in a thermopneumatic fluid and sequentially activated from left to right. The flow channel is fabricated by a proprietary etching technique to closely match the contour of the bulging silicon nitride membrane. Heating the fluid deflects the membrane that displaces the liquid. The deflected membrane seals the channel to prevent backflow. Performance of the micropump has been predicted to be 7 μL per minute at 15 psi. Pumps that can displace precise amounts of liquid have applications in medicine for automatic insulin dispensing, as well as in manufacturing for precise process control and reduced material cost.

(a)

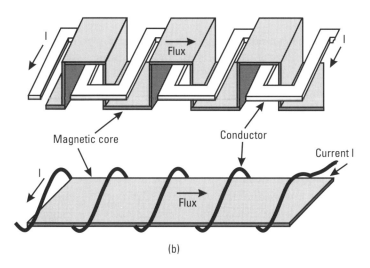

(b)

Figure 9.3 (a) Magnetic micromotor, and (b) multilevel meander coil. (Courtesy of Georgia Institute of Technology.)

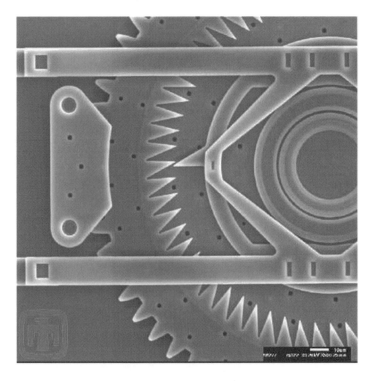

Figure 9.4 Rotary wedge stepper motor. (Courtesy of Sandia National Laboratories.)

9.2.4 Microdynamometers

The first steps have been taken to achieve a functional planar microdynamometer [9]. A dynamometer consists of a motor, a coupling gear train, a generator to act as an active load, and associated electronics. The LIGA process (discussed in Chapter 2) with the addition of a sacrificial layer (SLIGA) was used to fabricate mechanical components of the microdynamometer. Magnetic actuation was chosen for the micromotor and generator windings. As shown in Figure 9.6, a two-pole-pair motor with three windings per pole (upper right corner) and a generator with six windings were fabricated using electroplated nickel. Photodiodes integrated into the design can be used to determine the position of the rotor in the motor and the generator. Among the issues that must be resolved to realize a functional microdynamometer are magnetic materials problems.

A microtransmission fabricated using surface micromachining can increase the power from a microengine by a factor of 3 million when the friction is neglected [10]. Researchers coupled six identical transmission systems, each with a 12:1 reduction ratio to achieve the total increase. That level could

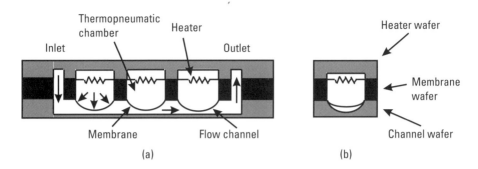

Figure 9.5 Three thermopneumatic actuators providing peristaltic pumping action: (a) side view and (b) end view.

Figure 9.6 Microdynamometer fabricated with electroplated nickel. A single-pole-pair test drive is at the bottom right. (Courtesy of University of Wisconsin.)

allow a micromachine the force needed to move a 1-lb object. The gear is reversible and can increase as well as decrease speeds. Figure 9.7 shows an example of a surface micromachined gear set [11].

Figure 9.7 Four gears forming a transmission for increasing the force from a microengine. (Courtesy of Sandia National Laboratories.)

9.2.5 Microsteam Engines

The world's smallest steam engine was fabricated with surface micromachining technology [11]. Figure 9.8 shows the SEM of the device. Water inside the compression cylinders is heated by electric current, vaporizes, and then pushes the piston out. Capillary forces retract the piston when the current is removed.

9.2.6 Actuators in Other Semiconductor Materials

Actuators have been designed and fabricated using thin film and plated metals, dielectrics, and photoresists for sacrificial layers on GaAs and InP substrates. GaAs and InP materials are used to fabricate MMIC devices, which require tuning that potentially can be performed by on-chip actuators and result in improved performance and yield. Sliding interdigitated capacitive tuners, bending beams, and rotating switches have been fabricated in a MMIC-compatible process as a first step toward their ultimate use to control the MMIC [12].

Epitaxial silicon carbide (3C-SiC) films and sputtered amorphous SiC films have been investigated for high-temperature MEMS devices [13].

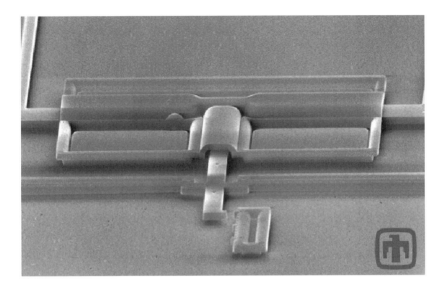

Figure 9.8 A single-piston microsteam engine. (Courtesy of Sandia National Laboratories [11].)

Suspended diaphragms and freestanding cantilever structures were etched into epitaxial films using bulk micromachining. The cantilever beams deflected downward due to residual stress variation in the film. Surface micromachining techniques were also applied to amorphous SiC. A 150-μm-diameter, 1.5-μm-thick gear was fabricated from amorphous SiC sputtered on silicon dioxide. Those materials hold promise for MEMS devices but require considerable development effort to progress from the laboratory into production.

9.3   Other Micromachined Structures

In addition to actuators, MEMS structures will be used for system components that require small size or reproducibility that can be achieved with micromachining. Multiple, metal microgears have been driven by forced air or relatively weak magnetic fields. Other examples discussed in this section demonstrate the variety of research and development activity that is occurring for MEMS-based devices.

9.3.1   Cooling Channels

A microheat pipe created by parallel microchannels in the bottom of a high-performance IC can provide cooling to minimize hot spots, improve performance, and increase reliability. A proposed approach is shown in Figure 9.9 [14]. After the channels are etched, multiple metallization layers are vacuum deposited to line the walls of the channel and seal the top. Heating the chip in a fluid bath fills approximately 20% of the cavity volume with fluid. The ends are sealed after this step to contain the fluid. The microheat pipe's operation causes fluid to evaporate in high-temperature regions and condense in low-temperature areas, resulting in a more uniform temperature distribution across the IC. Increasingly higher operating frequencies for higher performance MPUs and MCUs may require this type of cooling to avoid increasing the packaging size or the amount of external heatsinking.

A patent has been issued in the United States for the cooling of high-power semiconductor devices using microchannels formed by micromachining or laser cutting techniques in silicon or silicon carbide structures [15]. The microchannels remove heat by forced convection or the use of fluid coolant located as close as possible to the heat source. The microchannels maximize heatsink surface area and provide improved heat transfer coefficients for higher power density of semiconductor devices without increasing junction temperature or decreasing reliability.

Gaseous flow in micron-sized channels is also being investigated. The exchange of energy and momentum between MEMS and their gaseous environments often governs the response characteristics of these systems. To study the momentum exchange between a gaseous medium and microdevices, microchannels were fabricated (52.25 mm wide, 1.33 μm deep, and 7,500 mm long) for conducting nitrogen flow experiments [16].

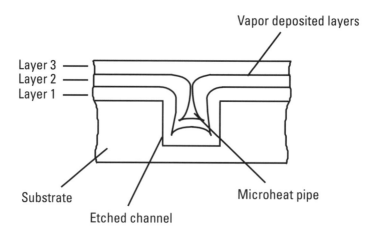

Figure 9.9 A microheat pipe for cooling integrated circuits.

9.3.2 Microoptics

A number of microoptical devices have been produced by micromachining techniques, including gratings, lenses, air bridges, electrical interconnects, fiber-optic couplers, alignment aids, corner reflectors, and waveguides [17]. In addition to silicon, a number of III-V semiconductor materials are being investigated for improved optical performance. A chemically assisted ion beam etching (CAIBE) process using argon and xenon ion beams and chlorine as the reactive gas was used to fabricate surface-emitting lasers. The 3-μm diameter lasers operated at room temperature with a threshold current below 1.5 mA and a differential quantum efficiency of 16%. However, a number of problems must be solved for the lasers to provide a viable solution for optical computing, chip-to-chip communications, and optical switching. The combination of mechanics, optics, and electronics at the micron scale promises to be an important field in optical microsystems.

A different combination of micromachining and optics is demonstrated in Figure 9.10 [18]. The polysilicon microscanner in the figure consists of a

Figure 9.10 A rotating polygon optical microscanner made by electroless plating of nickel. (Courtesy of Case Western Reserve University.)

hollow nickel-plated polygon reflector on the rotor of an electrostatic drive micromotor. A microoptomechanical system (MOMS) is based on wafer level integration of optical and MEMS components. Microscanners and movable optical elements have been designed and fabricated using electroless plating of reflective nickel surfaces on the rotor of the micromotor. The thickness (height) of the nickel is 20 μm, the width 10 μm. Scanners with diameters from 250 to 1,000 μm have been produced, but the larger micromotors do not operate reliably after they are released. Diffraction grating microscanners were fabricated with spatial gratings of 2 and 4 μm using a similar process. The key to MOMS devices is the ability to fabricate different optical and mechanical structures on a common substrate.

An actuator has been proposed that uses optical power to provide mechanical energy [19]. An optical actuator has potential advantages of higher operating speed, lower power consumption, and lower thermal expansion than nonoptical approaches. A silicon cantilever reacts to a photoelectric current by relaxing, as opposed to an applied electrostatic voltage, which stress the beam further.

9.3.3 Microgrippers

Several researchers have demonstrated microgrippers or microtweezers. One research team has developed a surface-micromachined polysilicon microgripper that is activated by an electrostatic comb-drive [20]. The electrostatic comb drive technique provides the force for several microactuators as well as an oscillating structure for many sensors. The two movable gripper arms are controlled by a three-element electrostatic comb, as shown in Figure 9.11. The length of the drive arms, L_{dr}, is 400 μm, and the length of each extension arm, L_{ext}, is 100 μm. By using separate open and close drivers, the gripping range for a given maximum voltage is doubled. Movement of 5 μm was produced by the grippers with less than 30V applied to the comb drive.

9.3.4 Microprobes

A cantilever beam contacting and scanning across the surface of a sample can measure the topography of a surface. The instrument, an atomic force microscope (AFM), requires a mechanical structure with a sharp tip, small spring constant, and high resonant frequency [21]. Batch fabrication yields cantilevers with very reproducible characteristics, and piezoelectric, capacitive, or piezoresistive sensing techniques can be used to sense the deflection of the probe tip. The construction of an AFM probe is illustrated in Figure 9.12(a). The dimensions of one device are $L1 = 175$ μm, $L2 = 75$ μm, $w = 20$ μm, $b = 90$ μm, and

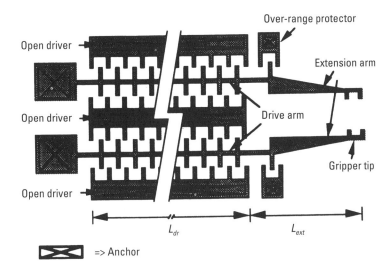

Figure 9.11 Schematic of a microgripper with an electrostatic comb drive. (*After:* [20].)

$t = 2\,\mu$m. The calculated spring constant of this structure is 4 N/m. The AFM probe was used to measure a silicon dioxide grating that had a depth of 270 Å and repeated every 6.5 μm. These types of devices may be useful in profilometry and IC inspection. An AFM probe produced by Park Scientific Instruments is shown in Figure 9.12(b).

9.3.5 Micromirrors

Micromachined digital micromirror devices can be used for displays. As shown in Figure 9.13, the micromirror element is an aluminum mirror suspended over an air gap by two thin, post-supported hinges [22]. The mechanically compliant hinges permit the mirror to rotate 10 degrees in either direction. The posts provide the connections to a bias/reset bus that connects all the mirrors of the arrays to a bond pad. The mirrors are fabricated over conventional CMOS SRAM cells that provide an address circuit for each mirror. The mirrors have a response time of approximately 10 ms and can be pulse-width modulated to provide a gray-scale output in a black-and-white display. Monolithic arrays with 768 by 576 pixels have been produced.

A hinged micromirror device has been fabricated using surface micromachining [11]. The hinged micromirror is driven by a microengine coupled to a three-gear torque-increasing system. The combination is sufficient to

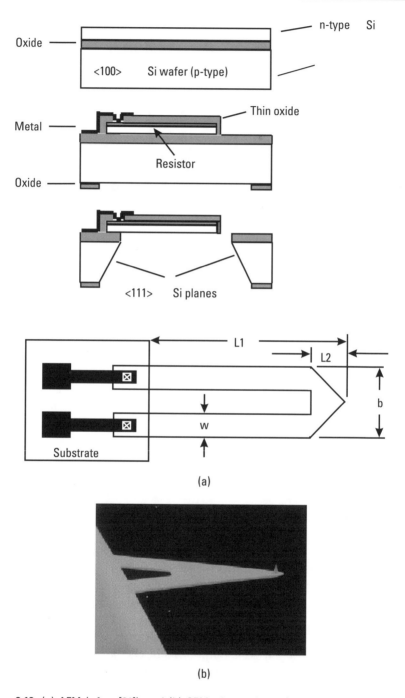

Figure 9.12 (a) AFM (*after:* [21]), and (b) SEM of actual unit (courtesy of Park Scientific Instruments).

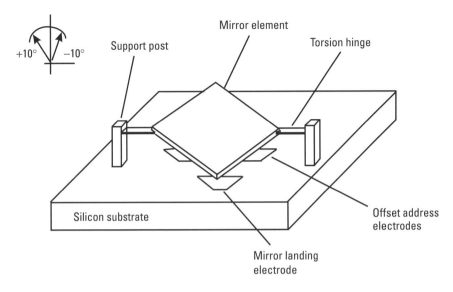

Figure 9.13 Digital micromirror with torsion hinge suspension.

deflect the mirror without aid of external probes. Figure 9.14 shows the mirror in the upright position.

9.3.6 Heating Elements

Multijunction thermal converters (MJTC) have been fabricated using a standard CMOS process and bulk micromachining [23]. After CMOS processing, a cavity is etched by bulk micromachining that produces a suspended MJTC cantilever structure, as shown in Figure 9.15. Polysilicon resistive heating elements and aluminum-polysilicon thermocouple junctions are encapsulated in glass. The glass protects those elements from the etchant and provides a mechanical support. MJTCs have potential applications in low-cost, high-precision RF and microwave power circuits.

9.3.7 Thermionic Emitters

Arrays of sputtered tungsten thermionic emitters have been fabricated by surface micromachining [24]. An SiO_2 layer isolates sputtered tungsten from the silicon substrate. The tungsten is patterned by wet etching, and the silicon is also wet-etched under the filament to avoid contact. The filaments were tested in a vacuum of $5 \cdot 10^{-7}$ torr. The filaments changed from barely visible red to white light once they start to emit. Operating life approaching 1 hr has been

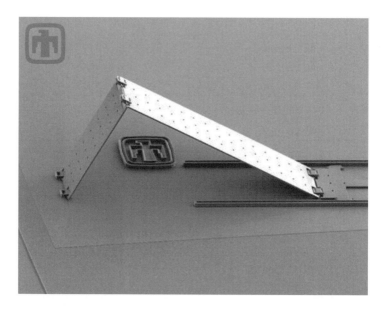

Figure 9.14 Hinged micromirror showing hinge and rail assemblies. (Courtesy of Sandia National Laboratories.)

Figure 9.15 An MJTC. (Courtesy of NIST and Ballantine Laboratories, Inc.)

achieved with emission currents of about 10 nA. The thermionic emitters are used as the first stage on a miniature SEM. The SEM is less than 2 cm³ in volume and is formed by stacking five silicon dice.

Those recent developments in thermionics require examining terms that are commonly used for semiconductors. The invention and volume production of semiconductors or solid-state devices displaced vacuum tubes. However, thermionic devices made using semiconductor processes have now created gaseous-state microdevices. Volume production of such devices is certain to raise comments if they are referred to as solid-state devices.

9.3.8 Field Emission Displays

A number of field emission displays (FEDs) are being developed using micromachining techniques [25]. An FED consists of an array of emitting microtips. In one approach, the microtips are formed by etching chambers inside a masked silicon wafer. Molybdenum tips are vacuum deposited in the chamber. Several hundred emitter tips are fabricated for each pixel, allowing dozens of tips to fail without discernible loss of brightness.

Another technique to produce the microtips uses selective etching of a polycrystalline silicon substrate. This self-aligned process uses the crystal structure of silicon to produce atomically sharp silicon tips. It may be possible to eliminate lithography in this approach due to the self-aligning nature of the process. (The application of FEDs is discussed further in Chapter 11.)

9.3.9 Unfoldable Microelements

Microstructures that have elastic joints have been used to make movable three-dimensional structures from planar surface micromachined structures. Microrobots are among the possibilities for such devices. The basis of the movable structure is shown in Figure 9.16(a) [26]. Polyimide provides the flexible connection to polysilicon plates or skeletonlike structures. The PSG is sacrificially etched to free the structure for the substrate. Released structures have actually been folded at the hinges, like paper, creating microcubes and insectlike three-dimensional structures.

An SEM of a robotic bug example is shown in Figure 9.16(b) [27]. The structure was fabricated using MCNC's multiuser MEMS process (MUMPS). The bug includes an actuator, electronics, sensors, an IR diode, and a solar cell. The microrobot is made from a silicon chip 1 cm by 0.5 cm by 0.5 mm.

A surface-micromachined MEMS hinge designed for foldout structures uses two structural layers of polysilicon to form both the hinge and the movable plate [28]. The most recent development is an improvement over an earlier

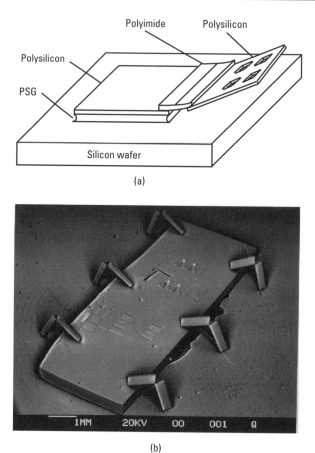

Figure 9.16 (a) Three-dimensional structure with elastic joint (*after:* [27]), and (b) robotic
bug (courtesy of BSAC, University of California, Berkeley).

pin-and-staple system and scissor-hinge design. The pin-and-staple system
required three structural layers to achieve a base, axle pin, and retaining staple
where the foldout movable structure was attached to the substrate. The scissor-
hinge weaves a second polysilicon layer under and then over the first polysilicon
layer. Weaving allows the retaining structure to be deposited with the first layer
but not connected to it. The most recent development makes use of cantilevers
to press on and then fix the axle pin. The axle is designed with a self-
limiting mechanism that prevents the axle from protruding beneath the first
layer of polysilicon. The hinge was designed, built, and tested in a surface-
micro-machined microphotonic system. The hinge could be opened to 90 ±0.8
degrees.

Methods for actuating the artificial arms on robotic devices have been studied and compared to properties of nature's actuators: muscles [29]. One of the more interesting developments is an electrostatic muscle that uses the combination of a number of small force displacements to produce significant displacement [30]. Integrated force arrays produce a flexible membrane that contracts with an applied electric field. Surface micromachining is used to create metal-clad polyimide rectangles that are only a few millimeters on a side. An array of 1.5 million cells forms a 1-cm-long fiber that contracts 0.3 cm with only a few volts applied. The force-array technology is being developed for coordinated movement of interlocking flexible elements. Besides robotics, self-aligned high-density multicontact electrical connectors can be designed using this technology.

9.3.10 Micronozzles

Microminiature apertures and nozzles are required for optical instruments and a variety of mechanical devices, including high-resolution ink jet printers, flow control, and atomizers [31]. The micronozzle used for ink jet printers is the highest volume production microactuator. Sacrificial etching is used to produce a highly cusped nozzle-shaped structure using silicon nitride. A mold is created for the nitride by steps (a) through (d) in Figure 9.17. The nitride structure is

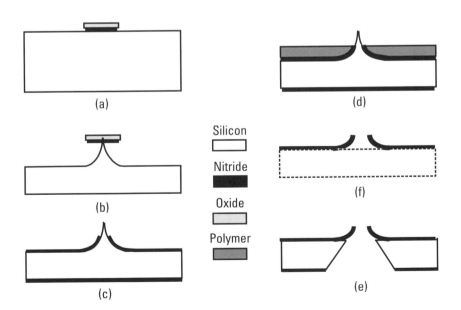

Figure 9.17 Micronozzle fabrication process.

freed from the substrate by a KOH etch and precision backsawing, as shown in Figure 9.17(e). Alternatively, back-masked anisotropic etching produces the structure shown in Figure 9.17(f). A variety of other materials, including silicon dioxide, boron-doped silicon, polysilicon, refractory, and noble metals, can be used for the nozzle.

9.3.11 Interconnects for Stacked Wafers

Wafer-on-wafer construction is being investigated to improve the density of integrated circuits. Micromachining with subsequent metalization can provide a technique to interconnect a stack of various silicon technologies at the wafer level. As shown in Figure 9.18 [32], anisotropically etched wafers are aligned where the interconnection is required. The pyramid-shaped structure has a square opening at the top of 25 mm and is 120 mm at the bottom for thin wafers. A fine gold-plated wire mesh is used to fill the cavity. The wire mushrooms at the top and compresses at the bottom to form a gold-on-gold contact.

9.3.12 Nanoguitar

In case the reader's imagination has not been sufficiently stimulated by the variety of structures discussed so far, a final example is the nanoguitar fabricated by researchers at Cornell University [33]. As shown in Figure 9.19, what has been called the world's smallest guitar is only 10 μm long. The six strings are each 50 nm, or 100 atoms, wide. The nanoguitar was micromachined from crystalline silicon. The key to the technology is electron-beam (E-beam) lithography,

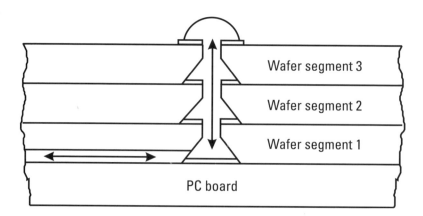

Figure 9.18 Interconnects from pyramid-shaped vias for multiple layer structure. (*After:* [32].)

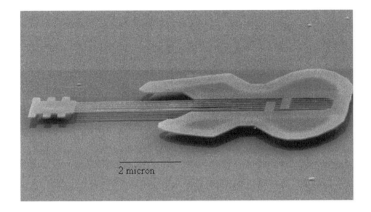

Figure 9.19 Nanoguitar fabricated using E-beam lithography.

which allows extremely fine patterns to be created. A 30-nm-wide wire suspended across a 2-μm gap has also been made using this approach. These nanomechanical devices are among the initial research to resolve physical questions regarding motion and mechanical energy dissipation.

9.4 Summary

Micromachining technology is used to produce micron-scale machines—movable structures—that are independently fascinating. If they can be interconnected for use in a microsystem, such devices have the potential to create new industries just as semiconductor technology has done. MEMS devices combined with sensors will provide new tools and improved performance for control systems.

References

[1]   Bryzek, J., and J. R. Mallon, "Silicon Integrated Circuit Sensors and Actuators," *Wescon Professional Advancement Program Session 9*, Nov. 14–15, 1989, San Francisco, pp. 196–201.

[2]   Koeneman, P., et al., "Feasibility of Micro Power Supplies for MEMS," *J. Micromechanical Systems*, Vol. 6, No. 4, Dec. 1997, pp. 355–362.

[3]   Zdeblick, M., "A Revolutionary Actuator for Microstructures," *Sensors*, Feb. 1993, pp. 26–33.

[4] Leonard, M., "Electric Motors on a Chip Advance From Academia's Labs," *Electronic Design*, Jan. 25, 1990, p. 26.

[5] Allen, M. G., "Polyimide-Based Processes for the Fabrication of Thick Electroplated Microstructures," *7th Internat'l Conf. on Solid State Sensors and Actuators (Transducers '93)*, June 7–10, 1993, pp. 60–63.

[6] Ahn, C. H., and M. G. Allen, "A Fully Integrated Micromagnetic Actuator With a Multilevel Meander Magnetic Core," *Technical Digest IEEE Solid-State Sensor and Actuator Workshop*, June 22–25, 1992, Hilton Head, SC, pp. 14–18.

[7] Allen, J. J., and H. K. Schriner, "Micromachine Wedge Stepping Motor," *1998 ASME Internat'l Mechanical Engineering Congress and Exposition,* Nov. 15–20, 1998, Anaheim, CA.

[8] Folta, J. A., N. F. Riley, and E. W. Hee, "Design, Fabrication and Testing of a Miniature Peristaltic Membrane Pump," *Technical Digest IEEE Solid-State Sensor and Actuator Workshop*, June 22–25, 1992, Hilton Head, SC, pp. 186–189.

[9] Christenson, T. R., et al., "Preliminary Results for a Planar Microdynamometer," *IEEE 91CH2817-5 Transducers '91*, pp. 6–9.

[10] "Microtransmission Gives 3 Million-to-One Gear Ratio in a Square Millimeter," *Machine Design*, Feb.19, 1998.

[11] Sandia MEMS http://www.mdl.sandia.gov/Micromachine/

[12] Hackett, R. H., L. E. Larson, and M. A. Melendes, "The Integration of Micro-Machine Fabrication With Electronic Device Fabrication on III-V Semiconductor Materials," *IEEE 91CH2817-5 Transducers '91*, pp. 51–54.

[13] Tong, L., M. Mehregany, and L. G. Matus, "Silicon Carbide as a Micromechanics Material," *Technical Digest IEEE Solid-State Sensor and Actuator Workshop*, June 22–25, 1992, Hilton Head, SC, pp. 198–201.

[14] Markstein, H., "Embedded Micro Heat Pipes Cool Chips," *Electronic Packaging and Production*, Oct. 1993, p. 14.

[15] Hamilton, R. E., et. al, "Microchannel Cooling of High Power Semiconductor Devices," U.S. Patent No. 5,801,442, Issued Sept. 1, 1998.

[16] http://goesser.mit.edu/MTL/Report94/MEMS/gaseous.html

[17] Deimel, P. P., "Micromachining Processes and Structures in Micro-Optics and Optoelectronics," *J. Micromechanics and Microengineering*, Dec. 1991, pp. 199–222.

[18] Merat, F., and M. Mehregany, "Integrated Micro-Optical-Mechanical Systems," *Proc. SPIE*, Vol. 2383, Feb. 1995.

[19] Tabib-Azar, M., "Optically Controlled Silicon Microactuators," *Nanotechnology*, 1990, pp. 81–92.

[20] Kim, C.-J., et al., "Polysilicon Microgripper," *Technical Digest IEEE Solid-State Sensor and Actuator Workshop*, June 4–7, 1990, Hilton Head, SC, pp. 49–51.

[21] Tortonese, M., et al., "Atomic Force Microscopy Using Piezoresistive Cantilever," *IEEE 91CH2817-5 from Transducers '91*, pp. 448–451.

[22] Mignardi, M. A., "Digital Micromirror Array for Projection TV," *Solid State Technology*, July 1994, pp. 63–68.

[23] Gaitan, M., J. Kinard, and D. X. Huang, "Performance of Commercial CMOS Foundry Compatible Multijunction Thermal Converter," *7th Internat'l Conf. on Solid State Sensors and Actuators (Transducers '93)*, June 7–10, 1993, pp. 1012–1014.

[24] Perng, D. C., D. A. Crewe, and A. D. Feinerman, "Micromachined Thermionic Emitters," *J. Micromechanics and Microengineering*, Vol. 2, No. 1, Mar. 1992, pp. 25–30.

[25] Derbyshire, K. "Beyond AMLCDs: Field Emission Displays?" *Solid State Technology*, Nov. 1994, pp. 55–65.

[26] Shimoyama, I., et al., "Insect-Like Microrobots With External Skeletons," *IEEE Control Systems*, Feb. 1993, pp. 37–41.

[27] http://www-bsac.eecs.berkeley.edu/~yeh/sems.html

[28] Friedberger, A., and R. S. Muller, "Improved Surface Micromachined Hinges for Fold-Out Structures," *J. Microelectromechanical Systems*, Vol. 7, No. 3, Sept. 1998, pp. 315–319.

[29] Hunter, I. W., and S. Fafontaine, "A Comparison of Muscle With Artificial Actuators," *Technical Digest IEEE Solid-State Sensor and Actuator Workshop*, June 22–25, 1992, Hilton Head, SC, pp. 178–185.

[30] Brown, C., "Force Arrays Mimic Natural Motion," *Electronic Engineering Times*, June 20, 1994, pp. 41, 49.

[31] Farooqui, M. M., and A. G. R. Evans, "Microfabrication of Submicron Nozzles in Silicon Nitride," *J. Microelectromechanical Systems*, Vol. 1, No. 2, June 1992, pp. 86–88.

[32] Markstein, H., "Vertical Wafer Integration Optimizes Memory Density," *Electronic Packaging & Production*, Jan. 1995, p. 30.

[33] http://www.news.cornell.edu/science/July97/guitar.ltb.htn

10

Packaging, Testing, and Reliability Implications of Smarter Sensors

I regret that there are no more worlds to conquer.
—Alexander the Great

10.1 Introduction

All the advances and research that are occurring in micromachining would lead one to believe that breakthroughs in those areas will be sufficient to revolutionize sensing. Unfortunately, the problems associated with the basic sensor packaging are compounded when the sensor is combined with higher levels of electronics. These problems initiate at the lowest level of die and wire bonding and extend to encapsulation, sealing, and lead-forming issues. Fundamental assembly differences frequently exist between sensor and microelectronics packaging and are among the problems that must be solved to achieve smarter sensors. These differences include die bonding for stress isolation instead of for heat dissipation and wire bonding procedures.

Packaging is essential to establishing the reliability of the sensor. Therefore the reliability requirements must be taken into account in the design of the package, especially for custom packages in specific applications. Testing of the sensor-circuitry combination also requires combining test capability from both technologies. This chapter addresses sensor packaging technology, especially new technology from the semiconductor industry, that should be applied to smart sensors, reliability, and testing concerns for smarter sensors.

10.2 Semiconductor Packaging Applied to Sensors

Many of today's sensor packages resemble semiconductor packages of the 1980s or even the 1970s. The semiconductor industry has made significant progress in high-density plastic encapsulated packages. The increased use of surface-mount technology (SMT) is among the more important changes. To achieve increased functionality without increased silicon complexity, available silicon technologies are being combined at the package level in packages based on semiconductor, not module manufacturer, assembly techniques. These multichip modules (MCMs) are being evaluated for several applications, including automotive.

As the use of surface-mount technology increases, a decline is occurring in the use of the previously popular dual in-line plastic (DIP) package. Other through-hole packages like single in-line plastic (SIP) and pin grid array (PGA) will also not increase. New SMT approaches, like ball grid array (BGA) and microBGA packages are the focus of current packaging development. For future, highly integrated components, packaging techniques must take into account more complex, system-level requirements, as well as SMT assembly requirements. The ability of the sensor industry to adapt to the newer semiconductor packaging approaches to sensors will determine the acceptance of smart sensor technology and the future growth of the industry.

The packaging, assembly, and testing contribution to the cost of a well-established micromachined pressure in a DIP package is shown in Figure 10.1(a) [1]; the sensor chip is only 14% of the total cost. For more complex packaging, such as a pressure sensor, in which media isolation is required, the cost of packaging, assembly, and testing can climb to 95% of the total cost.

The type of sensor, the amount of circuitry, and the application can change the cost significantly. An accelerometer has a separate control chip and g-cell in a surface-mount package. The die cost in that case is 30% for the g-cell and 24% for the ASIC, as shown in Figure 10.1(b). However, the greatest challenge with the accelerometer for automotive air bag applications is inexpensive testing [1].

Sensor packages have basic requirements that are similar to those of semiconductor devices. The variety of harsh sensor applications makes packaging more difficult than packaging for a semiconductor device. However, the basic package operations occur in similar order.

A completed sensor wafer has a final processing step that prepares it for packaging [2]. That step could include thinning the wafer and attaching a backside metal such as a gold-silicon eutectic. Sensors tested at the wafer level that do not meet minimum specifications are identified as rejected units by an ink dot. Sensors are then separated into individual dice from the wafer by

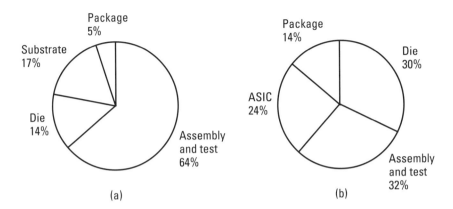

Figure 10.1 Cost of the elements of (a) a pressure sensor in a DIP package compared to (b) an accelerometer in a surface-mount package. (*After:* [1].)

sawing or scribe-and-break techniques. Good sensor dice are placed in carriers that allow automatic pick and place machines to transfer dice from the carrier to the final package, where a die bond attaches the sensor firmly to the package. Wire bonds connect the electrical contacts on the die surface to the leads of the package, which allow the sensor to interface to external components. The package is then sealed, if it is metal or ceramic, or encapsulated, if it is molded plastic. Lead plating and trim (or singulation) occur next and are followed by marking and final test operations.

The actual design of the package must take into account sensitive areas of the semiconductor device and the sensor's specific function. The following list [2] includes the characteristics that affect sensor packaging. The sensitivity of the semiconductor to light must be minimized in packaging for an accelerometer but optimized for a photodiode. Similarly, during design the package's sensitivity to stress must be taken into account to prevent stress from affecting the offset and sensitivity in a stress-sensitive pressure sensor. For smart sensors, the key item in the list is the fact that integration level affects the sensor's package.

- Wafer thickness and wafer stack (e.g., single, silicon-silicon, silicon-glass);
- Dimensions;
- Environmental sensitivity/requirement for physical interface;
- Physical vulnerability/stress sensitivity;
- Heat generation;
- Heat sensitivity;
- Light sensitivity;

- Magnetic sensitivity;
- Integration level.

A passivation layer deposited near the end of the wafer fabrication process protects the active area of semiconductor devices. Table 10.1 [2] lists the common terms used to describe that layer. Silicon dioxide and silicon nitride doped with boron, phosphorus, or both are two materials used for the passivation process. For the semiconductor sensor, the mechanical properties of the layers also must be taken into account. For example, in silicon pressure sensors, implanted or diffused elements are protected by a passivation layer, but the diaphragm area is masked to avoid the dissimilar material interface and dampening effect that would be caused by the glass layer.

For semiconductor components, the package provides protection from environmental factors, which can include moisture and gaseous or liquid chemicals. The packages are further protected in applications like automotive underhood-mounted modules by additional epoxy and silica potting compounds or conformal coatings (e.g., acrylics, polyurethane, silicone, and ultraviolet curing compounds) that cover the printed circuit to which the component is mounted. However, semiconductor sensors frequently have to interface to the environment. Pressure sensors, for example, that respond to static and dynamic pressures must have protection techniques that allow the pressure signal to be transmitted to the force-collecting diaphragm with minimal damping and distortion of the signal. Distortion can be caused by compressible fluids.

Table 10.1
Silicon Wafer Passivation Layer Terms (*After:* [2])

Material	Assembly Level
Silox	Wafer
Vapox	Wafer
Pyrox	Wafer
Glassivation layer	Wafer
PSG	Wafer
BSG	Wafer
PBSG (phospho-borosilicate glass)	Wafer
Parylene	Package
Dimethyl-silicone	Package

Additional protective materials for the die surface and the wire bonding connections, such as a compliant thin conformal parylene coating or hydrostatic methyl-silicone gel, are frequently used as the means to transmit pressure to the top surface of the sensor. Parylene deposition is a vacuum process in which the reactive vapor is passed over a room-temperature sensor and coats the sensor with the polymer [3]. The equipment used to perform the process is quite sophisticated, especially relative to gel coatings. The kind of media to which parylene- and gel-protected sensors can be interfaced is limited by the properties of the protective material.

10.2.1 Increased Pin Count

One of the more difficult problems that must be solved when additional electronic circuitry is integrated with or interfaced to the sensor is the requirement for additional pinouts. Integrated circuits, including MCUs, have industry-accepted standard packages that allow a large number of pinouts. To increase the density in rapidly increasing surface-mounted applications, pin pitches below 0.5 mm are being pursued. Tape-automated bonding (TAB) is one of the technologies being developed to address fine-pitch requirements. Sensors, on the other hand, usually are limited to eight or fewer pins. Furthermore, the packaging varies considerably from manufacturer to manufacturer, with no standard form factor (Figure 10.2). The requirements for a mechanical interface for pressure, force, flow, or liquid level cause additional packaging problems.

Additional circuitry, whether it is simple signal conditioning or more sophisticated approaches that include MCU capabilities, has an impact on sensor packaging [4]. A "sensor-only" versus integrated sensor plus control circuit is one level of differentiating packaging requirements. However, increased functionality through additional circuitry, either on the same chip or from a separate chip included in the ultimate sensor package or module, affects the pin count, normally increasing the number of package pins. An exception occurs for simple amplification and temperature compensation circuits, where the number of pinouts is actually reduced from four to three for piezoresistive sensors.

10.3 Hybrid Packaging

Hybrid packages, such as ceramic multichip packages, are routinely used during the research and development of semiconductors and MEMS devices. They allow researchers to probe selected portions of the die and verify expected functionality, especially if the entire die is not performing to predicted performance levels. Hybrid packages are also used for production sensors.

10.3.1 Ceramic Packaging and Ceramic Substrates

Ceramic packages, such as the ceramic DIP (CERDIP), utilize a leadframe that is attached to the ceramic base through a glass layer. After die and wire bonding, a ceramic top is glass-sealed to the base. The same technique is used for other form-factor ceramic packages, including the ceramic flat pack. Ceramic packages are usually used for high-reliability applications and are much more expensive than other semiconductor packaging techniques. They are very useful in the development phase of a sensor because the silicon die does not have to be encapsulated. That allows various test points on the die to be easily probed and measured in packaged form [5].

Ceramic technology is used in hybrid assembly techniques for sensors. The ceramic substrate, usually an aluminum oxide, provides a firm mounting platform for the sensor die. Stress isolation can be obtained by utilizing a compliant silicone for the die attachment. The ceramic substrate allows laser trimming of thick-film resistors deposited on the ceramic surface to provide the calibration for a signal-conditioned sensor.

10.3.2 Multichip Modules

The extensive research that is being performed for packaging multichip microelectronics is also being evaluated and adapted for manufacturing combined sensor(s) and microelectronics. Figure 10.3 shows potential packaging techniques that could possibly be used for sensors [4]. These approaches include (a) conventional chip-and-wire, (b) flip-chip, and (c) TAB. Chip-and-wire is the standard die-on-substrate packaging technique. Flip-chip packaging is discussed in Section 10.4.3. TAB packaging eliminates wiring bonding from the die to a lead by directly attaching a lead to the top of the die. Any one of these methods is a potential candidate for MCM packaging.

For MCMS with one or more sensing elements and silicon ICs, silicon is one of the substrate materials that could be used. The approach of a silicon circuit board to achieve wafer-level, hybrid integration may be the best interim solution for a combined sensor and MCU. However, there are several varieties of MCMs, depending mainly on the substrate technology: MCM-L, MCM-C, MCM-D, and MCM-Si are common classifications [6].

An MCM-L (the -L suffix stands for *laminate*) uses advanced printed circuit board technology, copper conductors, and plastic laminate–based dielectrics. It is essentially a chip-on-board (COB) technology, as shown in

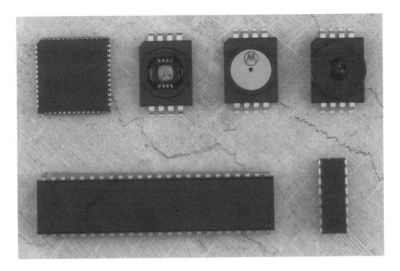

Figure 10.2 Pressure sensor and accelerometer packages compared to microcontroller packages.

Figure 10.3(a). An MCM-C uses thick-film screening on cofired ceramic substrates. An MCM-D has interconnections formed by the thin-film deposition of metals on dielectrics of polymers or inorganic compounds. An MCM-Si uses a silicon substrate with aluminum or copper interconnections and SiO_2 as the inorganic dielectric media.

The yields for MCMs can be lower than for assemblies made using packaged semiconductors. Recent known good die (KGD) efforts in the semiconductor industry are directed toward improving yields by improved testing at the wafer level. The same approaches will be required for MCM sensors.

10.3.3 Dual-Chip Packaging

Custom packages that accommodate a sensor die and its associated signal conditioning circuit are common for high-volume applications. In those instances, the volume justifies the development and tooling costs. An example of a pressure sensor package developed for automotive manifold absolute pressure sensors is shown in Figure 10.4 [7]. Thin-film resistors on the control die are laser trimmed to provide a fully signal conditioned and calibrated pressure sensor. The control die has been used as a building block for other pressure applications and with different packaging requirements. The two-chip approach has proved to be both cost effective and flexible.

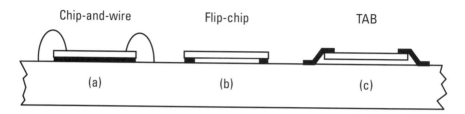

Figure 10.3 Bare die mounting techniques for MCMs.

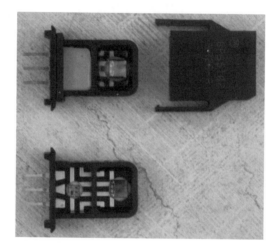

Figure 10.4 Dual-chip custom plastic package. (Courtesy of Motorola, Inc.)

10.3.4 Ball Grid Array Packaging

Ball grid array (BGA) packages are among the newer semiconductor packages that may also have potential for multichip smart sensors. An overmolded pad array carrier (OMPAC™) package is one of the new developments in semiconductor packaging.[1] It is also known as solder bump array (SBA), pad array carrier (PAC), and land grid array (LGA). Each technique relies on a blind solder joint for direct attachment of a leadless chip carrier to a printed circuit board [8]. As shown in Figure 10.5, the OMPAC™ has solder bumps attached to the copper foil pad on an epoxy glass substrate. The die in the plastic molded package is attached by gold wire bonds. Thermal vias run between solder bumps to

---

1. OMPAC™ is a trademark of Motorola, Inc.

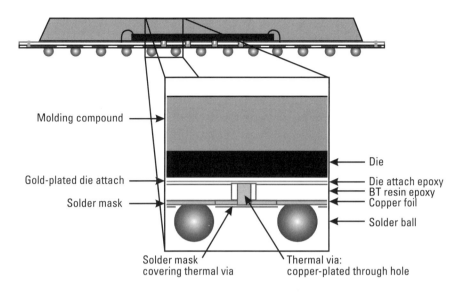

Molding compound

Gold-plated die attach

Solder mask

Die

Die attach epoxy
BT resin epoxy
Copper foil

Solder ball

Solder mask
covering thermal via

Thermal via:
copper-plated through hole

Figure 10.5 Plastic BGA package.

dissipate heat from the package. The solder bumps reflow during the surface-mount assembly process to attach the package to a printed circuit board.

10.4 Packaging for Monolithic Sensors

The highest volume semiconductor packaging techniques utilize molded plastic packages with form factors that include SIPs, DIPs, quad flat packs, and a variety of surface-mount devices (SMDs). The leadframe and molding techniques used in plastic packages provide the lowest cost semiconductor packages. Recent improvements include improved capability to withstand autoclave testing for improved hermeticity and the use of improved, low-stress mold compounds for surface-mount applications. However, a sensor that must physically interface with a mechanical system, such as a pressure sensor, presents additional packaging challenges.

Cylindrical TO-5 and TO-8 metal-can semiconductor packages have been adapted for usage as sensor packages. For pressure sensors, the lid and the header of the package usually have a hole to allow the pressure source to be applied to the die and provide a pressure reference to atmosphere. The leads extend through the base of the package through glass mounting seals. The use of gold plating material and hard die attachment to the glass substrate of the sensor provides one of the most media-compatible sensors. The metal lid is welded to the header, and pressure sensor port attachment is accomplished through solder, brazed, or welded techniques [5]. Such packages are also used for accelerometers, optical devices, and other sensors.

10.4.1 Plastic Packaging

A plastic chip carrier package that uses leadframe technology has been used for manufacturing pressure sensors for over a decade. A single mold forms both the body and the back of the chip carrier. The patented unibody package provides lower cost, fewer process steps, higher pressure range capability, and greater media compatibility compared to earlier versions made of separate body and metal back plate [5]. The leadframe assembly technique allows easy handling of several devices at one time and the automation of assembly operations such as die bond, wire bond, and gel-filling operations. Automation allows tight process controls to be implemented and still provides high throughput.

One of the more difficult problems for sensors that interface to harsh environments is media compatibility. Figure 10.6 is one approach to achieving a low-cost interface.

10.4.2 Surface-Mount Packaging

The packaging shift to SMT (see Figure 10.1) is also affecting sensors. That can mean specifically designed surface-mounted sensor packages or, in more complex sensors, a printed circuit board (PCB) assembly that uses SMT exclusively. An example of the extent of surface-mount packaging in a sensor is demonstrated by the GPS sensor shown in Figure 10.7. That 2-in by 3.25-in by 0.5-in PCB has five surface-mounted ICs, including a 32-bit MCU [9].

The single-board radar sensor in Figure 10.8 shows several surface-mounted components [10]. As volume increases in those applications, increased integration and possibly higher density MCMs will allow smaller

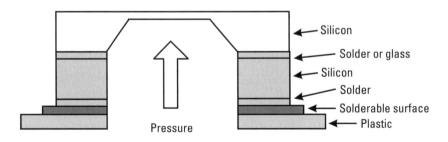

Figure 10.6 Increased media compatibility pressure sensor packaging.

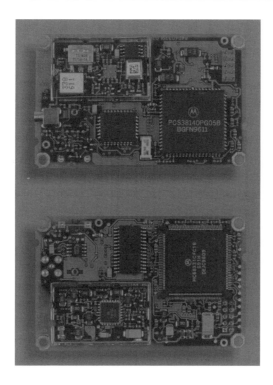

Figure 10.7 Surface-mount packaging in Oncore™ GPS. (Courtesy of Motorola, Inc.) (Oncore™ is a trademark of Motorola, Inc.)

packaging. High-volume manufacturing cost for a 0.5-in^2 version of the circuit in Figure 10.8 is estimated to be about $1.

10.4.3 Flip-Chip

Flip-chip packaging technology, commonly used in ICs, is starting to receive attention in sensors [11]. Flip-chip technology will allow the electronics to be fabricated on one chip, which can be attached to the MEMS chip through either a fluxless process or flux-assisted solder reflow process. This is an MCM-Si approach to combining different silicon processing technologies instead of increasing masking layers and die size to achieve a monolithic silicon solution.

Figure 10.9(a) is an example of flip-chip technology in a standard IC process [12]. The technology has been used in automotive and computer applications for several years. The solder bump is formed over plated copper at the wafer level. Several steps precede the actual formation of the bump. A

Figure 10.8 Indoor radar sensor. (Courtesy of Lawrence Livermore National Laboratory.)

passivation nitride is used to mask areas other than the aluminum-copper metal contact on the silicon die. Sputtered titanium-tungsten (TiW) and copper are used to form a base metal for copper, lead, and tin, which are plated onto the base. Reflow in a gradually ramped oven causes a thermally induced amalgam of tin and lead to form into a ball due to surface tension. Spacing of the bumps evenly around the die avoids stress during thermal expansion.

Figure 10.9 also shows how flip-chip technology can be applied to a sensor. The sensor is represented by a diaphragm structure but could be any micromachined structure. In one case, the sensor could be a flip-chip on a more complex IC such as an MCU. An alternative approach is to have the IC as a flip-chip on the micromachined structure. The requirements of the sensor must be considered when this type of packaging is developed.

10.4.4 Wafer-Level Packaging

Chapter 2 discussed surface-micromachined structures sealed inside a two-layer bulk micromachined sensor package. That wafer-level "packaging" of micromachined sensors results in considerable cost savings over packaging methods that use metal can or ceramic packages. The process has demonstrated

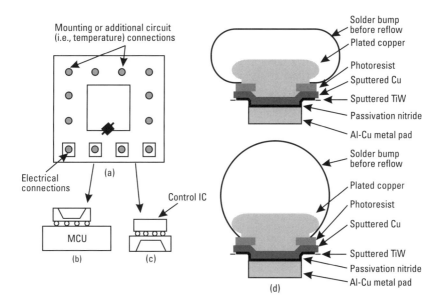

Figure 10.9 Flip-chip assembly: (a) die with bumps, (b) bumped sensor on MCU, (c) bumped control die on sensor, and (d) bump structure itself.

manufacturability and yields a well-protected, hermetic sensor device such as an accelerometer. The technology has far-reaching implications and opens the door to new methods of low-cost, hermetic packaging over the more expensive methods used today. Protection of the two-layer structure has been accomplished with a molded plastic package for current products. Future protective packaging may take a different form.

Epoxy protection for ICs has provided a low-cost packaging solution for consumer electronics. The glob-top package shown in Figure 10.10(a) is simply an epoxy mound that protects the die surface and wire bonds for an IC mounted on a circuit board. For flip-chips, a protective layer around the die has been proposed, as shown in Figure 10.10(b) [13]. In that case, the protection is from silicon to silicon with sensitive layers sealed inside the epoxy. Lower cost packaging will be essential for many new smart sensor applications. Adapting the approaches described here could provide the cost breakthrough.

10.5 Reliability Implications

Sensors require tests similar to those performed on ICs and unique qualification tests to verify that they will have acceptable performance of both the

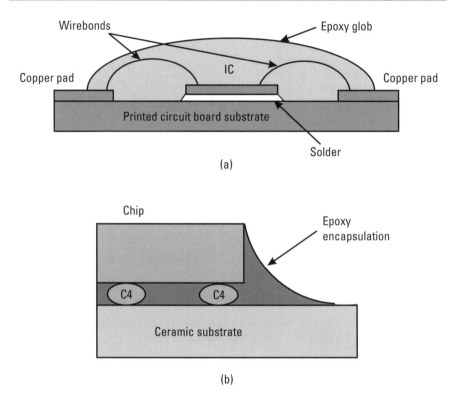

Figure 10.10 Epoxy protection for single die: (a) glob top and (b) flip-chip. (*After:* [13].)

silicon and the packaging for their intended applications. For example, several tests have been developed for silicon pressure sensors based on the need to detect potential failures due to the environment in which the devices will operate. The tests have been used to qualify sensors with on-chip amplification. Key tests include but are not limited to the following [14]:

- Operational life, such as pulse, pressure-temperature cycling with bias;
- High humidity, high temperature with bias;
- High temperature with bias;
- High and low temperature storage life;
- Temperature cycling;
- Mechanical shock;
- Variable frequency vibration;

- Solderability;
- Backside blow-off (for pressure sensors);
- Salt atmosphere.

Those tests use accelerated life and mechanical integrity testing to determine the lifetime reliability statistics for silicon pressure sensors. Potential failure mechanisms are determined by the materials, processes, and process variability that can occur in the manufacturing of a particular sensor. For example, the pressure sensor in Figure 10.11 has 10 product-related and 8 process-related areas with 73 different items that can affect reliability [14]. Other sensors will have reliability test requirements commensurate with the application for which the sensor is being used and the type of packaging technique that is employed.

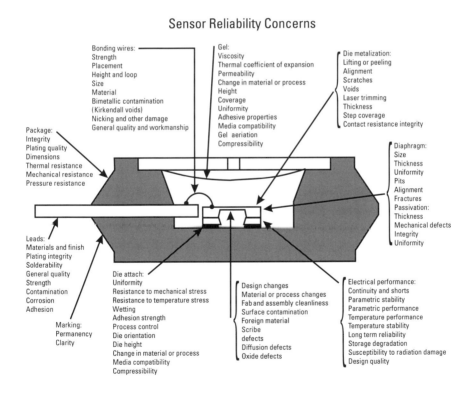

Figure 10.11 Reliability areas in a plastic packaged pressure sensor.

The newest approach for establishing media compatibility for sensors uses in situ monitoring [15]. A typical three-step approach for verifying media compatibility is to (1) characterize, (2) expose to environment, and (3) retest for conformity to specification. That requires removing the sensor from the test environment and also changes the exposure of the sensor through the addition of oxygen prior to retest. New procedures have been developed that test the organic compounds used in the sensor's packaging, the silicon sensing element, and the integrated circuitry in the sensor, without removing the sensor from the test chamber. New sensor materials and modified packaging designs have been shown to survive in a variety of harsh media common to applications in automobiles and trucks. Previously a comparable confidence level would have been possible only after achieving millions of unit hours in the actual application or experiencing a high rate of warranty returns.

10.5.1 The Physics of Failure

The sensor packaging problems specific to the application must be thoroughly addressed to ensure acceptable performance over the sensor's expected lifetime. An approach has been proposed to improve the reliability and reduce the cost of sensors [3]. Identifying minimum expectations and critical application factors that limit device lifetime are part of a methodology known as the *physics of failure* approach to reliability testing. The physics of failure approach involves analyzing the potential failure mechanisms and modes. Once those are understood, application-specific packaging development can proceed, the proper test conditions can be established, and critical parameters can be measured and monitored [16].

One key failure mechanism for sensors that operate in harsh media applications (e.g., pressure sensors) is corrosion, specifically, galvanic corrosion caused by dissimilar metals in electrical contact in an aqueous solution. Using a physics of failure approach, one can start to ask the appropriate questions. Which environmental factors contribute to the failure mechanism? What accelerates it? What should be done to the sensor package to prevent corrosion or minimize its impact over the sensor's lifetime? What is the expected sensor lifetime?

The answers to those questions allow simulations to be performed and costs to be analyzed before lengthy and potentially expensive testing is initiated. The testing will ultimately prove the acceptability of a particular proposal. Unless the proper mechanisms are known, overdesigning can add unnecessary cost to the packaging, such as a passivation layer, a hard die attachment, or any nonstandard process. Furthermore, the design tradeoffs that result may not prove beneficial. For example, adding a hard die attach to the pressure sensor to

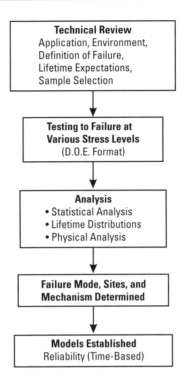

Figure 10.12 Physics of failure methodology.

improve the strength of the die bond could limit device performance in other areas, such as temperature coefficient of offset [16].

10.5.2 Wafer-Level Sensor Reliability

One of the inherent advantages of semiconductor sensors is their capability of integrating the signal conditioning, temperature compensation, and essentially any aspect of semiconductor fabrication on the same silicon substrate as the sensor. That allows wafer-level reliability to be implemented. Test structures on the wafer or monitor wafers allow evaluation of critical process steps as they occur and provide rapid feedback to detect and correct process errors. The cost of testing at the wafer level is considerably less than testing at the completed assembly level. As shown in Table 10.2 [17], the cost of evaluating critical process steps early in the wafer fabrication process provides a considerable cost reduction over testing performed after the assembly process is complete. The data is a comparison of semiconductor process testing only. However, as sensor technology advances, package approaches common in the semiconductor

Table 10.2
Cost of Testing for Various Tests (*After:* [17])

Test	Test Technique		
	Wafer level	Traditional	
		Normal	End of life
Gate-oxide evaluation	$200	$20,000	$32,000
Top-passivation evaluation	$175	$12,000	$16,000
New etch process	$800	$15,000	$13,000
Electromigration testing	$350	$7,000	$35,000
Hot-electron susceptibility	$750	$9,000	$8,500

industry, such as flip-chip and KGD, will also be applied to sensors. Wafer-level reliability will have increased interest for both the sensor manufacturer and the user.

Normal test techniques require packaged units and considerably more time for evaluation. End-of-life techniques require even longer testing to ensure that a product life (e.g., 20 years) is achievable. For wafer-level testing, additional testing at the package level is required because packaging methods have a significant effect on the total reliability. However, wafer-level reliability testing is one of the ways that semiconductor sensors can continue to achieve lower product cost and continuous improvement.

10.6   Testing Smarter Sensors

Several issues occur for testing sensors that combine other semiconductor technology. Because the ultimate smart sensor (level V in Figure 1.7) is a monolithic combination of a micromachined structure and an MCU, the challenges that that combination poses for testing must be addressed. Those challenges include, but are not limited to, different types of testing required, unique test equipment, and multipass testing. To allow for a low-cost integrated solution, those issues must be addressed [18].

The integration of a physical measurement function onto an already complex mixed-mode analog-digital chip such as an MCU with an integral ADC raises the need for an additional type of testing. If a pressure sensor is combined with an MCU, one of the more difficult testing situations occurs. Testing for the pressure sensor must be able to apply the physical medium being tested to the device and measure the response. Measuring the response to a physical stimulus is not a standard test for the semiconductor industry, especially under multiple temperatures. Standard testers can test the digital and analog portions of the MCU. The application of a physical stimulus and the ability to heat and cool the device under test rapidly and accurately drive the need for a modified and unique tester. Because the testers are one of a kind and not available as standard equipment, the cost of the tester is a large part of the final unit's cost.

Not only are the testers expensive, but the throughput is limited. Limited throughput will cause increased cost to each part due to increased depreciation costs allocated to each device. The cost is further increased by the need for multipass testing. For calibration, each part may require testing at least three different temperatures, to determine the transducer's output characteristics over temperature. Those values are then used to derive the compensation algorithm that is loaded into the on-chip EPROM. To complete the cycle, the device is again tested over temperature to confirm the accuracy. Therefore, not only is a special tester required, but it becomes a bottleneck because it must be used twice to complete each device—once to measure the characteristics and a second time to verify the result.

Finding ways to reduce the cost of testing is one of the key issues to make a low-cost integrated sensor and MCU a reality. Ideas that could prove promising include a thorough characterization of the design, limiting the operating temperature, limiting the accuracy, programming the MCU to take data during testing, and loading the test and compensation algorithm into the MCU prior to testing.

10.7 Summary

As smarter sensing technology is developed, microcontroller or ASIC capabilities will be included with the sensor. That will occur either as a monolithic structure or as multichip components; in either case, an increased number of pins will be added to the other sensor packaging requirements. However, all the justification for future cost effectiveness of semiconductor-based sensors is predicated on those sensors following the learning curve similar to that of semiconductors. For sensors to approach the progress made in semiconductors, plastic packaging must be developed. Furthermore, packaging standards must be standardized for test equipment (handlers) and second sources. Sensor packaging draws heavily from both semiconductor and hybrid packages, but new techniques that are unique to smart sensors most likely will be developed to meet future requirements.

References

[1] Vaganov, V., H. Joseph, and S. Terry, "Matching Customer Application Requirements With Cost Effective Sensor Packaging and Testing," *Proc. Sensors Expo*, Boston, May 13–15, 1997, pp. 181–189.

[2] Van Zant, P. *Microchip Fabrication*, New York: McGraw-Hill, 1990.

[3] Benson, A. F., "Count on Conformal Coatings," *Assembly Engineering*, June 1990, pp. 20–23.

[4] Frank, R., and J. Staller, "The Merging of Micromachining and Microelectronics," *Proc. Third Internat'l Forum on ASIC and Transducer Technology*, Banff, Alberta, Canada, May 20–23, 1990, pp. 53–60.

[5] Ristic, L. J., *Sensor Technology and Devices*, Norwood, MA: Artech House, 1994.

[6] Blood, W. R., and J. S. Carey, "Practical Manufacturing of MCMs," *Surface Mount Technology*, Nov. 1992, pp. 16–22.

[7] Czarnocki, W. S., and J. P. Schuster, "Robust, Modular, Integrated Pressure Sensor," *Sensor 95 Internat'l Conference on Sensors, Transducers, and Systems*, Nuremburg, Germany, May 1995.

[8] Chin, S., "Ball Grid Array Package Challenges Quad Flatpack," *Electronic Products*, Apr. 1993, pp. 19–20.

[9] Oncore Ad, *Automotive Industries*, Oct. 1994, p. 71.

[10] Babyak, R. J., "Electronics," *Appliance Manufacturer*, May 1994, pp. 95–99.

[11] Markus, K. W., V. Dhuler, and A. Cowen, "Smart MEMS: Flip Chip Integration of MEMS and Electronics," *Proc. Sensors Expo*, Cleveland, Sept. 20–22, 1994, pp. 559–564.

[12] *Linear/Interface ICs Device Data Book*, Vol. II, DL128/D Rev. 4, Motorola Semiconductor Products Sector, Phoenix, 1994.

[13] Hill, G., J. Clementi, and J. Palomaki, "Epoxy Encapsulation Improves Flip Chip Bonding," *Electronic Packaging and Production*, Aug. 1993, pp. 46–49.

[14] Maudie, T., and B. Tucker, "Reliability Issues for Silicon Pressure Sensors," *Proc. Sensors Expo '91*, Chicago, Oct. 1–3, 1991, pp. 101-1 to 101-8.

[15] Frank, R., and T. Maudie, "Surviving the Automotive Environment," *Electronic Engineering Times 1995 Systems Design Guide*, Special Issue, 1995.

[16] Maudie, T., D. J. Monk, and R. Frank, "Packaging Considerations for Predictable Lifetime Sensors," *Proc. Sensors Expo*, Boston, May 13–15, 1997, pp. 167–172.

[17] Reedholm, J., and T. Turner, "Wafer-Level Reliability," *Microelectronics Manufacturing Technology*, Apr. 1991, pp. 28–32.

[18] Frank, R., and D. Zehrbach, "Testing the System on a Chip," *Proc. Sensors Expo*, San Jose, CA, May 19–21, 1998, pp. 235–238.

11

Mechatronics and Sensing Systems

People are fascinated by what they can't quite understand.
—Michael Palin of Monty Python's Flying Circus

11.1 Introduction

To achieve the highest level of smart sensing, the entire system essentially has to become a sensing system or an extension of sensing capability. The term *mechatronics* is frequently used to describe subsystem design for a combination of electromechanical components. Computers or MCUs are an integral part of many mechatronic systems. Mechatronics has been defined as the synergistic combination of precision mechanical engineering, electronic control, and the systems approach for designing products and manufacturing processes [1]. Mechatronics represents a higher level of sophistication and complexity, requiring designing at the interface of the electromechanical system. In some ways, mechatronics is similar to analog-to-digital interface, only more complex. Although engineers from different disciplines share their results on a particular program, they work with a less-than-perfect understanding of how their design decisions affect the decisions of their counterparts. Mechatronics and other approaches, such as concurrent engineering, are attempting to understand design tradeoffs earlier in the design process to reduce design cycle time and ensure that the final product meets the design criteria. Mechatronics combined with micromachining and smart sensors will enable complete microsystems to be designed and fabricated in a single monolithic chip.

11.1.1 Integration and Mechatronics

The need for increased control capability in electronic systems is driving increased levels of integration in semiconductor electronics, including sensors. A fully integrated silicon system is frequently cited as a goal for the ultimate system. Silicon integration is certainly progressing toward that capability, especially for selected portions of systems. However, the cost and performance capabilities of higher levels of integration must compete against the components that they would displace. Also, design flexibility and use in several applications must be considered, including such diverse requirements as satisfying both domestic and international requirements, simultaneously meeting cost-sensitive, low-end specifications, and providing feature-sensitive high-end products. Furthermore, safety and reliability concerns, such as the potential failure modes from a system failure-mode-and-effect analysis (FMEA) can lead to more distributed intelligence when single-point failures cannot be tolerated. In some cases, the system constraints make it desirable to have sensing and power control elements remote—or at least separated—from the MCU that is controlling the system. That allows improved power dissipation, fault detection, and protection from system voltage extremes [2].

System partitioning and the determination of how much integration should occur within each component are decisions that should be made by an interactive design team, composed of both system and silicon designers. Integration for the output section of the control system is frequently called a smart-power IC. Those integrated solutions have been in volume production by several manufacturers since the mid-1980s. Understanding their contribution to the control system and how they take advantage of embedded sensing in both normal operation and fault modes is essential to achieve the systems approach and implement mechatronics methodology. This chapter examines smart-power ICs, more complex sensing in arrays, and systems aspects of mechatronics.

11.2 Smart-Power ICs

Smart-power ICs and power ICs (PICs) typically combine bipolar and MOS circuitry with power MOSFET technology to provide direct interface between the MCU and system loads, such as solenoids, lamps, and motors. Smart-power ICs are defined by the Joint Electron Devices Engineering Council (JEDEC) as "hybrid or monolithic devices that are capable of being conduction cooled, perform signal conditioning, and include a power control function such as fault management and/or diagnostics. The scope [of this definition] shall apply to devices with a power dissipation of 2W or more, capable of operating at a case temperature of 100°C and with a continuous current of 1A or more." Smart-power devices also provide increased functionality as well as sophisticated diagnostics and protection circuitry. Sensing of current levels and junction temperature is a key aspect of the design for control during normal operation and for detecting several fault modes.

The power in smart-power ICs is typically a power MOSFET or a diffused MOS device (DMOS). The size of this portion of the silicon area is determined by the process technology, voltage rating, and desired on-resistance. Process technology determines the on-resistance per unit area and has provided improvements based on reducing the cell size and improving cell geometry. The voltage rating inversely affects efficiency: a higher voltage rating means a higher on-resistance for a given area for the power device and lower efficiency. For a given process and voltage rating, the number of cells is increased to meet on-resistance specification. Because increased silicon area directly affects the cost, the highest on-resistance to allow safe power dissipation in the system normally is specified.

The smart-power approach to system design means that a number of circuit elements can be consolidated into a single device. These devices previously would have been discrete components or a combination of a standard (or custom) IC and discrete output devices. As illustrated in Figure 11.1, input, frequently from an MCU, initiates the startup or turn-on procedure. The control can have circuitry for soft-start function or charge pump circuitry for increasing the gate drive voltage. A high degree of functionality is achieved by smart-power ICs, because of the interaction among the control circuit, the power device, the protection circuitry, and the load. Multiple power devices on a single chip is the most cost-effective use of this technology. That provides space saving, component reduction, total system cost reduction, improved performance, and increased reliability from the reduced number of interconnections.

Historically, the choice of process technology for smart-power ICs has depended on the type of control elements that were integrated. Some circuit elements, like op amps, comparators, and regulators are best implemented using a bipolar IC process. MOS circuitry handles logic, active filters (time delays), and current mirrors better than bipolar circuitry. Some circuits, such as ADCs or power amplifiers, can be implemented equally well in either technology. Today, a process that has both MOS and bipolar for the control circuitry does not have to sacrifice performance or features; if it is combined with the appropriate output devices, the process can handle the power control functions for a number of system loads. However, just as a smart sensor needs to fully utilize the capability of a more expensive process, a smart-power IC also must provide advantages beyond those that separate control IC and discrete power

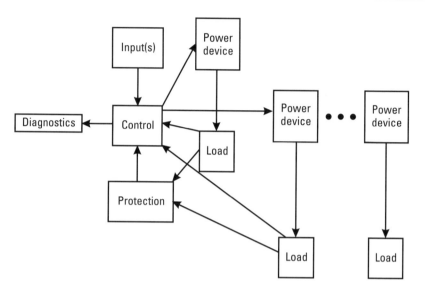

Figure 11.1 Smart-power IC block diagram.

devices can provide. One of the key means to achieving increased functionality is to integrate sensing into the smart-power IC.

11.3 Embedded Sensing

Sensing in smart-power ICs detects fault conditions and threshold conditions, which allows implementation of control strategies. Overtemperature, overcurrent, and overvoltage conditions can be sensed, and the smart-power IC can be designed to react or simply provide a signal to the MCU for system-safe operation. This section explores the different sensing techniques and applications of the sensed signal.

11.3.1 Temperature Sensing

Temperature sensors can be easily produced in semiconductor devices by using either the temperature characteristics of base-emitter voltage (V_{be}) or a diffused resistor. In addition, circuitry is easily integrated to produce a monolithic temperature sensor with an output that can be easily interfaced to an MCU. By using an embedded temperature sensor, additional features can be added to ICs. (Note: A sensor becomes an embedded item in a semiconductor product when it has a secondary instead of primary function.)

Sensing for fault conditions, such as a short circuit, is an integral part of many smart-power ICs. The ability to obtain temperature sensors in a mixed bipolar-CMOS-DMOS semiconductor process provides protection and diagnostics as part of the features of smart-power ICs. The primary function of the power IC is to provide a microcontroller-to-load interface for solenoids, lamps, and motors. In multiple-output devices, sensing the junction temperature of each device allows the status of each device to be input to the MCU, and, if necessary, the MCU can shut down a particular unit that has a fault condition [3].

In ICs, the temperature limit can be sensed by establishing a band gap reference. Two diodes' junctions are biased with different current densities, but the ratio is essentially constant over the operating temperature range (−40°C to 150°C). The following equation shows how the differential voltage (ΔV_{be}) is related to the current (I) and emitter area (A) of the respective transistors:

$$V_{be1} - V_{be2} = \Delta V_{be} = \left[k_B T / q\right] \cdot \ln\left(\frac{I1 / A1}{I2 / A2}\right) \qquad (11.1)$$

where k_B is Boltzman's constant ($1.38 \cdot 10^{-23}$ J/K), T is the temperature in K, and q is the charge of an electron ($1.6 \cdot 10^{-19}$ C).

At a targeted high temperature, the negative temperature coefficient of V_{be} equals the positive temperature coefficient voltage developed across a sense resistor, and a thermal limit signal is generated. Other circuitry regulates the voltage and turns off the output power device.

A smart-power IC can have multiple power drivers integrated on a single monolithic piece of silicon. Each driver can have a temperature sensor integrated to determine the proper operating status and shut off only a specific driver if a fault occurs. Figure 11.2 shows an eight-output driver that independently shuts down the output of a particular driver if its temperature is excessive (between 155 and 185°C).

The octal serial switch (OSS) adds thermal sensing through overtemperature-detection circuitry to the protection features. Faults can be detected for each output device and individual shutdown can be implemented. In a multiple-output power IC, it is highly desirable to shut down only the device that is experiencing a fault condition and not all the devices that are integrated on the power IC. With outputs in various physical locations on the chip, it is difficult to predict the thermal gradients that could occur in a fault situation. Local temperature sensing at each output, instead of a single global temperature sensor, is required.

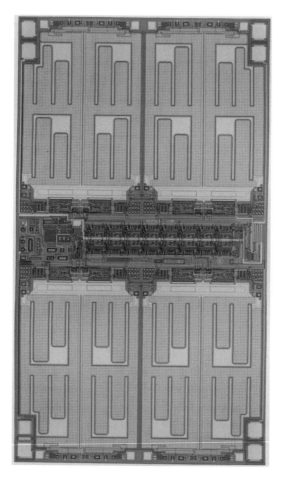

Figure 11.2 Photomicrograph of eight-output PIC. (Courtesy of Motorola, Inc.)

As shown in Figure 11.3, the eight outputs of the device have individual temperature sensors, which allow each device to turn off independently when the thermal limit of 170°C is exceeded. All the outputs were shorted to 16V supply at a room temperature ambient. A total current of almost 30A initially flowed through the device. Note that each device turns off independently, with the hottest device turning off first. Variations can result from differences in current level and thermal efficiencies. As each device turns off, the total power dissipation in the chip decreases and the devices that are still on heat up slower.

The hard short could have been diagnosed by short-circuit (current limit) sensing. However, a soft short, which by definition is below the current limit but exceeds the power-dissipating capability of the chip, can be an extremely

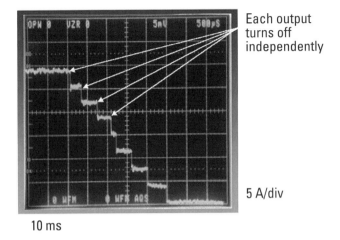

Figure 11.3 Independent thermal shutdown of an eight-output smart-power IC.

difficult condition to detect. Soft shorts require overtemperature sensing to protect the IC from destructive temperature levels.

The overtemperature condition sensed by the power IC may mean that the device turns itself off to prevent failure in one case; in another situation, a fault signal provides a warning to the MCU, but no action is taken, depending on the fault circuit design. The remaining portion of the system is allowed to function normally. With the fault conditions supplied to the MCU, an orderly system shutdown can be implemented.

Temperature sensing also has been integrated into a standard power MOSFET process. The power MOSFET process was chosen instead of a power IC process because a large die size was required to meet the power dissipation requirements of a very cost sensitive automotive application. A minor modification (the addition of a single masking layer) to a production power MOSFET process allows both sensing and other protection features to be integrated.

The thermal sensing is provided by integrated polysilicon diodes. By monitoring the output voltage when a constant current is passed through the diode(s), an accurate indication of the maximum die temperature can be obtained. A number of diodes are actually provided in the design. A single diode has a temperature coefficient of 1.90 mV/°C. Two or more diodes can be placed in series if a larger output is desired. For greater accuracy, the diodes can be trimmed during the wafer probe process by blowing fusible links made from polysilicon.

The response time of the temperature-sensing diodes is less than 100 μs, which has allowed the power device to withstand a direct short across an automobile battery. The sensor's output was applied to external circuitry that provided shutdown prior to device failure. The sensing capability also allows the output device to provide an indication (with additional external circuitry) if the heatsinking is not proper when the unit is installed in a module, or if a change occurs in the application that would ultimately cause a failure.

11.3.2 Current Sensing in Power ICs

Current sensing provides protection and control in power ICs. A device that demonstrates that approach is shown in Figure 11.4 [4]. The circuit, shown in Figure 11.4(a), includes a voltage regulator, an oscillator, and a charge pump to provide a high-side switch with an N-channel power MOSFET. This custom device has multiple current sensing levels for control and fault detection, as shown in Figure 11.4(b). The current-sensing technique uses a sample of the current flowing through the total power MOSFET cells, the ratio of the number of sample cells to control cells, as an indication of the total current. A resistor in series with the sense cells, which handles considerably lower current, provides a voltage proportional to the current in the power device. This SENSEFET™ technique avoids adding a series resistance to the total on-resistance and the subsequent power consumption, which could be excessive for an integrated circuit.[1] The isolated sensing cells for four different current-sensing levels are inside the four "lassoed" areas (left side, lower section in the cell area) in Figure 11.4(b). The highest current level could trigger an overcurrent fault, and the lower levels can be used for system-specific control points.

11.3.3 Diagnostics

The eight-output driver with serial communications shown in Figure 11.2 represents a high level of complexity for diagnostics. Control and sensing of the faults of individual output drivers could require over 24 connections for eight output drivers. By using the SPI protocol described in Chapter 6, 8-bit serial control of the output and independent diagnostics are possible with only three connections. The SPI operates at a frequency of up to 2 MHz and utilizes the low-voltage, high-speed CMOS capability in the SMARTMOS™ (Motorola's power IC) process.[2] With both the input and the output tied directly to the MCU, the unit acts as a closed-loop subsystem between the load and the MCU for each load. By using a daisy chain technique, several additional devices can

---

1. SENSEFET™ is a trademark of Semiconductor Components Industries (d.b.a. ON Semiconductor).
2. SMARTMOS™ is a trademark of Motorola, Inc.

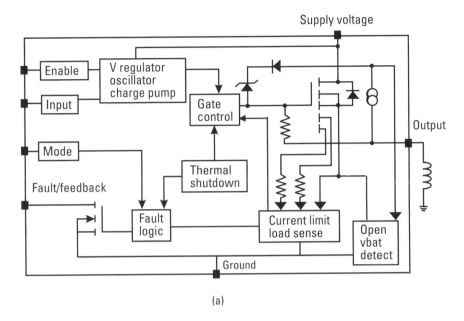

(a)

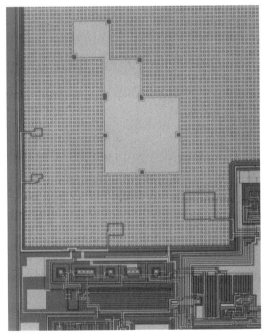

(b)

Figure 11.4 Current sensing integrated in smart-power IC: (a) block diagram, and (b) die photomicrograph. (Courtesy of Motorola, Inc.)

be added to the control capability. A total of 32 outputs are controlled with only four connections (the $\overline{\text{IRQ}}$, interrupt request not, is optional) to the MCU.

A high-level of smart-power IC device complexity is demonstrated in a parallel-serial control. The device has six outputs that can be selected independently on six dedicated parallel input pins by an input command (voltage above Vth) from an MCU. Inputs 0, 1, 4, and 5 also can be simultaneously selected. That allows the two output devices to be connected in parallel for reducing the on-resistance. Parallel control provides the fastest activation of the loads, which allows real-time control using PWM techniques. The drawback is that the highest number of connections must be made and the highest number of traces must be routed on the printed circuit board. Serial control of the six outputs is also possible using an SPI interface, as described earlier. Mixed control of the outputs can be achieved with various combinations of parallel and serial control of the outputs.

Fault diagnostics and control are similar to those in the OSS. However, an advanced SPI provides serial diagnostic information to the MCU, which includes immediate parity checking (nth word) to confirm that the smart-power IC received the word sent by the MCU. In the initial SPI implementation (the OSS), fault reporting is performed in the $n - 1$ word. The present command to the OSS is compared to the outputs from the previous word to determine the status. The time of the active filter period, which can be several clock cycles, is required before the next word can determine if a fault exists. The advanced SPI uses an exclusive OR to provide fault information when the current word is written.

Automotive applications have required diagnostics beyond those typically found in smart-power ICs. For example, the California Air Resource Board's latest OBDII legislation requires that the actual movement of the pintle in transmission solenoids be monitored when the output device is activated. That is done to indicate that the transmission is in the proper gear and that the torque converter has full lockup. Simply sensing the current going through the output device is insufficient to determine that the pintle has moved as a result of the output device being turned on. The differential sensing circuit in Figure 11.5(a) is used to detect the change that occurs when the solenoid's armature is attracted into the winding. The negative slope, shown in the current waveform in Figure 11.5(b), is an indication of the physical movement. Either low-side or high-side sensing can be performed with low-pass, high-pass filters and a comparator providing the signal conditioning. The event counter can be adjusted from 15 to 64 to establish the confidence level of detecting movement. The on-resistance of the power driver can be used as a sense resistor to convert the current into a voltage level. A peak detector also can be used to detect the peak in Figure 11.5(a). The results can be reported on a serial SPI or with parallel fault reporting. The OBDII diagnostic adds one more aspect to the fault-sensing/detection capability of smart-power ICs. Table 11.1 summarizes the faults that can be detected [2].

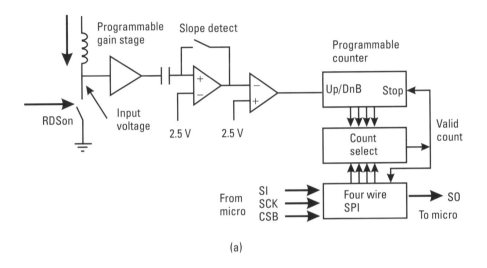

(a)

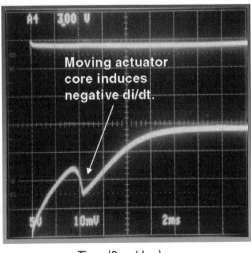

Time (2ms/dev)→

(b)

Figure 11.5 Detecting solenoid activation: (a) circuit, and (b) waveform.

Table 11.1
Fault Diagnostics for the Output

Fault	Technique
1. Open load (on-state)	Logic comparison
2. Open load (on- or off-state)	Logic comparison
3. Shorted load	Overcurrent sensing
4. Overvoltage (load dump)	Overvoltage sensing
5. Overtemperature (single output)	Temperature sensing
6. Independent overtemperature (multiple output)	Temperature sensing
7. Solenoid movement	Bipolar + logic

Fault-reporting techniques and corresponding MCU commands to the smart-power IC can be accomplished by several methods. For new devices, the most appropriate communication technique is dictated by the application requirements. Table 11.2 summarizes the communication techniques that are available as building blocks [2].

For distributed control systems, both smart-sensing and smart-power nodes are required. Smart-power ICs combined with an MCU or MCUs with integrated power devices implement the load control. That means the power or output side of the system must also be able to communicate using the system protocol. The protocols discussed in Chapter 6 must take into account the

Table 11.2
Summary of Communication Techniques for Power ICs

Parameter	Communication	Capability
Input	Parallel	Within slew rate (typ ≤ 1 A/μs)
	Serial	2-MHz SPI
	Parallel/serial	Selectable/mixed
Diagnostic	Flag (single)	Open/shorted load
	Flag (single)	Open (on- or off-state)/ shorted load
	Serial	2-MHz SPI (all faults)
	Flag (multiple with interrupt)	Reports all faults

power portion of the system. For a power node to be added to an existing system, it must have the appropriate protocol to operate on the system.

11.3.4 MEMS Relays

Electrostatic and electromagnetic MEMS relays have been produced using micromachining processes. Magnetically actuated relays can achieve larger forces and a greater air gap between contacts than electrostatic designs [5]. The larger gap provides greater isolation for applications at higher voltages. Micromachining processes used to produce the microrelay include surface micromachining and polyimide-mold electroplating techniques. The electromagnetic relays operate below 5V, so they can be directly interfaced to an MCU or a DSP. The relays have operated 850,000 cycles without failing. The combination of this adaptation of mechanical technology with micromachining and MCU and DSP technology should provide some useful mechatronic solutions.

11.4 Sensing Arrays

More than one sensor is frequently required to provide sufficient information for a control system. R&D efforts are progressing in several areas to integrate the sensors ultimately on the same silicon wafer with signal conditioning, computational capabilities, and possibly even actuation. In the short term, a single-package solution is proving to be an improvement over prior units. The sensing arrays can include a number of sensors for different measurands, for example, pressure, flow, temperature, and vibration. Multiple sensors of a given type can be used to increase the range, provide redundancy, or capture information at different spatial points. Also, multiple sensors of a given type, with minor modifications, can be used to measure different species in chemical sensing applications.

11.4.1 Multiple Sensing Devices

Improving the performance of sensors can be accomplished by simultaneously fabricating several sensors on the same silicon substrate or mounting several sensors in a hybrid package. One example of the monolithic approach is the surface micromachined pressure sensor developed by the Fraunhofer Institute of Microelectronic Circuits and Systems [6]. Surface micromachining was combined with CMOS processing to achieve a monolithic smart sensor. A single absolute capacitive surface micromachined pressure sensor with a

100-μm-diameter membrane provided approximately a 0.017-pF capacitance without applied pressure. Using up to 81 individual pressure sensors, which were switched in parallel, capacitance values between 1 and 2 pF were achieved. The sensor output varied approximately 0.2 pF over a 1- to 6-bar pressure range.

A multichannel probe designed for measuring single-unit activity in neural structures is shown in Figure 11.6 [7]. Eight active recording sites are selected from 32 sites on the probe shank using a static input channel selector. A large number of recording sites are desirable for sampling the total activity within even a restricted tissue volume. In this design, on-chip CMOS circuitry amplifies and multiplexes the recorded signals and electronically positions the recording sites with respect to the active neurons. The neuron signals typically have frequency components from about 100 Hz to 6 kHz, with an amplitude from 0 to 500 μV. The on-chip functions of the probe include channel selection, amplification, signal bandlimiting, multiplexing, clocking, power-on reset, and self-testing.

Multiple sensing sites are common in chemical measurements. A multielement smart gas analyzer has been developed that represents one of the most sophisticated sensors with respect to both operation and data processing [8]. The multielement sensor has a number of thin-film detectors to overcome the drawbacks of an earlier version and to improve selectivity and specificity in analyzing gas mixtures.

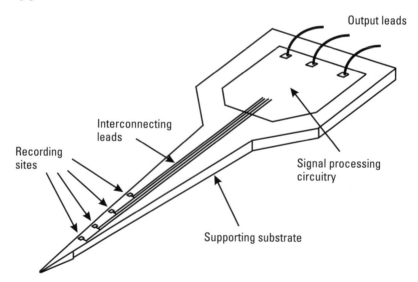

Figure 11.6 Multichannel microprobe sensor. (*After:* [8].)

A four-element gas analyzer is achieved by a three-chip hybrid configuration. Two dual-window gas detectors and a control interface chip are in the hybrid package. The gas detector process is compatible with CMOS processing so the combination eventually could be integrated into a single monolithic chip.

The control chip is capable of controlling four gas-detecting elements independently. All elements can be monitored and programmed simultaneously. A temperature of 1,000°C can be controlled for each chip. The chip communicates with a local-node MCU or a remote processor through an eight-pin standard interface. The front-end standard is capable of operating up to 32 sensors, 32 actuators, 32 self-test commands, and 32 special-function commands for different applications.

Another array approach for gas sensors has a 3-by-3 matrix of identical sensors on a single chip 4,466 by 6,755 μm [9]. The nine sensors are inter-digitated gate electrode field-effect transistors (IGEFETs). A single version of this design has selectively and reversibly detected 2 ppb of nitrogen dioxide and diisopropyl methylphosphonate, two pollutants that can adversely affect the nvironment. By using a Fourier transform signal processing technique and implementing a pattern recognition algorithm, it is possible for this type of sensor to operate as an electronic "nose" for detecting and identifying the constituents of a multicomponent gas mixture.

Photodiode arrays are becoming cost effective for manufacturing inspection, quality control, process monitoring, and other industrial applications. A silicon CCD has a spectral range from 200 to 1,100 nm. Devices can be manufactured with detector elements as small as several microns in width and with up to 2,048 elements. After amplification and digitizing for signal processing, the readout can routinely reach more than 1 MHz. A complete spectral measurement can be made in only a few milliseconds. Optical design, electronic design, and mechanical packaging must be coordinated in a mechatronics methodology to allow those devices to be used in a laboratory as well as a manufacturing environment [10].

An infrared focal plane array, which operates much like the human retina, has been designed and fabricated [11]. The array, called a neuromorphic sensor, has logarithmic sensitivity to avoid saturation common in conventional thermal imagers. It also performs pixel-based sensor fusion and real-time local contrast enhancement. The sensor contains a backside illuminated, 64-by-64 array of 100-μm indium antimony photovoltaic diodes. A CMOS IC performs the readout function as well as two-dimensional averaging across the chip. The neuromorphic sensor mimics the neural network process by interconnectivity between each pixel's four nearest neighbors. A digital computer performing the same functions at the same processing rate as the neuromorphic sensor would require approximately 100W. The neuromorphic device consumes only 50 mW, or nearly 1/2,000 of the power. That is a dramatic demonstration of the difference that smart sensors will contribute to future systems.

11.4.2 Multiple Types of Sensors

A variety of sensors have been fabricated on a single chip to detect damage and performance degradation caused during semiconductor assembly and packaging operations [12]. The sensors were designed using CMOS technology common to many IC manufacturing facilities. Sensors on the chip include ion detector, moisture sensor, electrostatic discharge detector, strain gauge for measuring shear forces, edge damage detector, corrosion detector, as well as heaters for corrosion, acceleration, and thermal modeling.

Figure 11.7 is another example of an integrated multiple-sensor chip that utilizes the pyroelectric and piezoelectric effects in zinc oxide thin films [13].

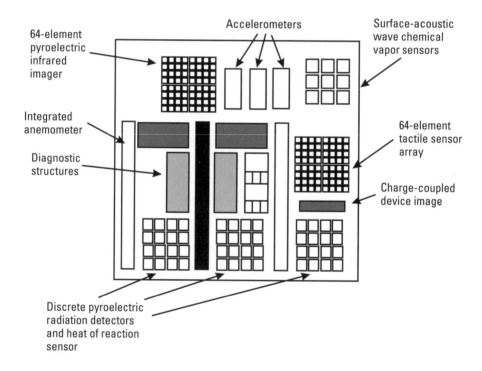

Figure 11.7 Multiple-type integrated sensor. (*After:* [13].)

Sensors on the chip include a gas flow sensor, an infrared-sensing array, a chemical-reaction sensor, cantilever beam accelerometers, SAW vapor sensors, a 64-element tactile sensor array, and an infrared CCD imager. The 8- by 9-mm^2 chip also has MOS devices for signal conditioning, array accessing, and output buffering. Backside micromachining of the silicon is performed as the last step in the fabrication process. The multielement sensor demonstrated low processing cost per function.

11.4.3 An Integrated Sensing System

An example of a fully integrated sensing system using multiple sensing elements is a monolithic hearing aid. The hearing aid requires only two functional elements to be a system: the sensor and the amplification. A micromachined piezoelectric microphone with on-chip CMOS signal conditioning has been designed and fabricated that has potential for hearing aid applications [14].

The sensor consists of eight piezoelectric (zinc oxide) electrode pairs. Bulk micromachining to define the microphone diaphragm is performed prior to CMOS processing. The design takes into account a 9-hour 1,150°C anneal and stresses imposed by subsequent CMOS processing steps. After CMOS circuitry is completed, the 0.5-μm thick zinc oxide is magnetron sputtered onto the diaphragm. The eight pairs of upper and lower electrodes are connected in two series groups of four electrode pairs. The microphone has a measured sensitivity of 0.92 mV/Pa and a resonant frequency of 18.3 kHz. Preamplifier noise of only 13 μV allows the microphone to achieve an A-weighted noise level of 57-dB sound pressure level (SPL), which is roughly the level of conversational speech.

11.5 Other System Aspects

The smart sensor system requires additional components to provide a self-contained unit. Current developments would allow batteries and displays to be integrated with the sensor. Any item mentioned in Chapters 8 and 9 might also be part of this integration—if the packaging problems discussed in Chapter 10 are solved. Those components, combined with smart sensors, digital logic from an MCU or a DSP, and smart-power ICs, will enable a number of new mechatronics approaches. One system aspect that must be addressed is the catastrophic effect that electrostatic discharge (ESD), system voltage transients, and EMI can have on electronic components. Those factors are covered in Section 11.5.3.

11.5.1 Batteries

Portable applications have generated considerable system design activity to obtain extended performance and improved life from the batteries that power these systems. By including semiconductor components in the battery, a smart battery is created. A smart battery is a battery or battery pack with specialized hardware that provides information regarding the state of charge and calculates predicted run time [15]. Information regarding the type of battery chemistry and the battery-pack voltage, capacity, and physical packaging is conveyed by a protocol to power-related devices in the system. The system can provide control to a smart battery or batteries, a smart battery charger, and various regulators and switches. Other battery-charging efforts are directed at eliminating the memory effect in nickel-cadmium (NiCd) batteries, reducing the charging cycle time, and avoiding overcharging, which reduces battery life. Portable data acquisition systems and remote sensing devices will benefit when smart batteries are incorporated into their design.

Reduced battery size through thin-film manufacturing technology will also affect sensors. A thin-film solid-state lithium-titanium-sulphur (Li/TiS$_2$) microbattery has been reported that has properties suitable for long-life rechargeable applications [16]. The microbatteries range in size from 8 to 12 μm in thickness and have a capacity between 35 and 100 μA-hr/cm^2. The open-circuit voltage is approximately 2.5V. Batteries have been cycled over 10,000 cycles at 100 μA/cm^2. Microbatteries have routinely withstood 1,000 cycles between 1.4V and 2.8V at current densities as high as 300 μA/cm^2.

Any reasonable smooth surface is a potential substrate for the thin-film battery. A chromium, TiS$_2$, and solid electrolyte layer are sputtered onto the substrate. LiI, Li, and a protective coating are vapor deposited over the sputtered layers. The construction technique will allow the microbattery to be incorporated with many semiconductor devices, including microsensors, during their manufacture.

Photopatterned carbon that is subsequently heat-treated at various temperatures has been explored for its potential in MEMS [17]. The material produced by this process permits a variety of new MEMS applications through new shapes, resistivities, and mechanical properties. Carbon surfaces can be used to form electrochemical electrodes with deposition of a wide variety of organic molecules. Polymeric batteries have been developed using the photopatterned carbon and surface micromachining.

11.5.2 Field Emission Displays

Efficient low-power displays are an essential part of many systems, including portable digital assistants, virtual reality–driven robots, and automotive GPSs.

The displays communicate information from the machine to the operator. Currently, active matrix LCDs are the dominant technology, but flat panel displays and an emerging technology called field emission displays are potential replacements in several areas. Field emission displays produce light using colored phosphors. They do not need complicated, power-consuming backlights and filters, and almost all light is visible to the user. In addition, no power is consumed by pixels in the off state. A field emission display consists of an array of microtips that are micromachined into a substrate. Metal, silicon, and diamond are being investigated for the microtips. One approach that uses MEMS technology and silicon as the backplane has been analyzed for its potential to be cost competitive versus other approaches [18]. The display is an array of field emitters on 0.9-μm centers used to provide a phase grating. The phase grating is operated by electrically controlling the mechanical positions of the grating elements to modulate the diffraction of light. The evaluation showed this MEMS approach to be promising because it uses a simpler process and half the MEMS-masking steps of digital micromirrors (see Section 9.3.5). Successful implementation of this or any other CMOS-based imaging approach being investigated for displays will provide additional possibilities for a sensor-to-operator interface.

11.5.3 System Voltage Transients, Electrostatic Discharge, and Electromagnetic Interference

Surviving voltage transients, ESD, and EMI are common application requirements, especially in the automotive environment. The ability to withstand a load dump is the prime factor in defining the voltage capability of semiconductor devices used in automotive applications. The load-dump transient is a high-voltage (>100V in an unsuppressed system), high-energy transient generated by disconnecting the battery from the alternator when the engine rpm is high and the alternator is generating a high-level output. Techniques of handling load dump include: (1) squelching the energy from an excessively high voltage input (>60V) by turning on an active clamp until the energy decays (few hundreds of milliseconds) and (2) designing the output devices to withstand high voltage without turning on. The latter approach is done with an increase in the resistance of each output device and therefore causes more power dissipation under normal operating conditions. That technique must be used for high-voltage (>100V) pulses to avoid excessive power dissipation during the load dump. A third alternative uses the circuitry in the smart-power device to turn the output on for a period of time (about 80 ms) when an overvoltage occurs. The input is time filtered before sampling to determine if the gate drain clamp should be turned on again [2].

Semiconductor sensor designs and applications have a system requirement that is normally not a problem in presemiconductor mechanical sensors, namely, ESD. Automotive and other applications generate high levels of ESD during manufacturing or in service. As a result, automotive manufacturers specify some of the most stringent ESD testing to qualify components. Automotive module manufacturers are indicating that future specifications will require any IC pin accessible outside the module to withstand 10-kV ESD with 250 pF and 1,500Ω load. That means peak currents approaching 7A and energy levels in the range of 12.5–14 mJ must be tolerated. The output connections of sensors, power, and control circuitry can be exposed to ESD during assembly or operation [2].

Sensors must be designed to meet levels of EMI commonly found in the application or required by industry specifications. EMI is a disturbance or malfunction of equipment or systems caused by the operation of other equipment or systems or by the forces of nature. Distributed control systems can be susceptible to high-frequency switching, which is common in computers, switching power supplies, and power control. Figure 11.8 shows an industrial process

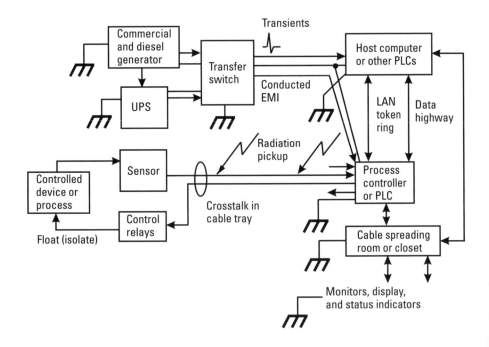

Figure 11.8 EMI in process control system. (*After:* [19].)

control with ground loops and other EMI paths [19]. Sensors easily can be affected, depending on the wiring. Isolating the wiring, avoiding ground loops, and in some cases using shielded wiring may be necessary to avoid problems.

In automotive applications, some manufacturers require the sensor to withstand electric fields of magnitude up to 200 V/m in the engine compartment and up to 50 V/m in the passenger compartment [20]. Performance levels and measurement methods are defined in a number of SAE documents, including SAE J551, SAE J1816, SAE J1113, SAE J1407, and SAE J1338. Classifications of failure-mode severity are provided in SAE J1812.

An example of automotive testing for EMI is demonstrated by the qualification testing for a plastic-packaged integrated pressure sensor. The testing required performance at 200 V/m radiated immunity from 10 kHz to 1 GHz. That is one of the higher signal strength levels specified in automotive standards. However, required performance depends on the customer specification. Over 30 EMI specifications have been reported for engine control applications around the world; there is no common standard [21]. The following field characteristics were defined:

- Field strength = 200 V/m;

- Frequency range = 1 kHz to 1 GHz;

- Modulation = AM, 30% with 1-kHz sine wave;

- Polarity = vertical and horizontal;

- Criteria for acceptable performance of device under test (DUT): alternating current (AC) ripple = ≤ ±50 mV.

Testing results included (1) passing the test and (2) finding that the test setup played a significant role in defining the device response and real susceptibility. Furthermore, due to the number of specifications that exist and to verify acceptable performance of product in the plastic package for two specific applications, additional tests were conducted. An extreme amount of EMI is encountered in a direct RF injection test. Direct RF injection is performed by soldering the harness to an impedance network. A 100-V/m signal was injected directly into the leads of the plastic packaged sensor. To pass the test, a filter technique was developed that greatly improved the performance of a plastic-packaged integrated pressure sensor even under the toughest testing, including the direct RF injection test.

11.6 Summary

This chapter examined the control of the output portion of the system made possible by smart-power ICs. That technology, combined with smart sensors and computing technology, is part of mechatronics, a holistic approach to the system. New developments in batteries and display technology will help to redefine the "sensor." Advances in all those areas will contribute to new generations of smarter sensing and smart sensing systems that are interactive sensing networks, especially if potential system problems are taken into account early in the design process. The complete microsystem is a promise that may be fulfilled in the future.

References

[1] Comerford, R. "Mecha ... What?" *IEEE Spectrum*, Vol. 31, No. 8, Aug. 1994, pp. 46–49.

[2] Wollschlager, R., K. M. Wellnitz, and R. Frank, "Diagnostics and Communications in Vehicle Control Systems Using Smart Power ICs," SAE SP-1013 *Sensors and Actuators 1994*, Warrendale, PA, pp. 21–27.

[3] Frank, R., "Embedded Temperature Sensors for Protection and Control," *Sensors*, May 1995, pp. 38–45.

[4] Himelick, J. M., J. R. Shreve, and G. A. West, "Smart Power for the 1990s in General Motors Automobiles," *ISATA 22nd Internat'l Symp. on Automotive Technology & Automation*, Florence, Italy, May 14–19, 1989.

[5] Mannion, P., "Batch Processing Yields High-Current, Low-Cost, Magnetically Operated Relays That Can Be Mounted on Circuit Boards," *Electronic Design*, Mar. 23, 1998, p. 40.

[6] Kandler, M., et al., "Smart CMOS Pressure Sensor," *Proc. 22nd Internat'l Symp. on Automotive Technology & Automation* #90185, Florence, Italy, May 14–18, 1990, pp. 445–449.

[7] Ji, J., and K. D. Wise, "An Implantable CMOS Analog Signal Processor for Multiplexed Microelectrode Recording Arrays," *IEEE Solid-State Sensor and Actuator Workshop*, Hilton Head, SC, June 4–7, 1990, pp. 107–110.

[8] Najafi, N. "A Multi-Element Gas Analyzer Utilized in a Smart Sensing System," *Proc. Sensors Expo West*, San Jose, CA, Mar. 2–4, 1993, pp. 97–109.

[9] Kolesar, E. S., "Sensitive and Selective Toxic Gas Detection Achieved With a Metal-Doped Phthalocyanine Semiconductor and the Interdigitated Gate Electrode Field-Effect Transistor (IGEFET)," *Proc. Sensors Expo*, Cleveland, Sept. 20–22, 1994, pp. 11–30.

[10] Pelnar, T., "Photodiode Array Spectra Analysis and Measurements for Industry," *Sensors Expo West*, San Jose, CA, Mar. 2–4, 1993, pp. 337–339.

[11] Massie, M. "Neuromorphic Sensor Mimics the Human Eye," *EDN Products Edition*, Jan. 17, 1994, p. 34.

[12] Yates, W., "Test Chips Detect Semiconductor Problems," *Electronic Products*, Sept. 1991, pp. 17–18.

[13] D. L. Polla, and R. S. Muller, "Integrated Multisensor Chip," *IEEE Electron Device Newsletter*, Vol. EDL-7, No. 4, Apr. 1986, pp. 254–256.

[14] Ried, R. P., et al., "Piezoelectric Microphone With On-Chip CMOS Circuits," *IEEE J. Microelectromechanical Systems*, Vol. 2, No. 3, Sept. 1993, pp. 111–120.

[15] Maliniak, D. "Smart Batteries Spelled Out by Joint Enabling Specification," *Electronic Design*, June 13, 1994, p. 50.

[16] Jones, S. D., and J. R. Akridge, "Microfabricated Solid-State Secondary Batteries for Microsensors," *Proc. Sensors Expo*, Cleveland, Sept. 20–22, 1994, pp. 215–223.

[17] Madou, M., et al., "Carbon Micromachining C-MEMS," *Proc. Symp. on Chemical and Biological Sensors and Analytical Electromechanical Methods*, Pennington, NJ, 1997, pp. 61–69.

[18] DeJule, R., "Flat Panel Technologies," *Semiconductor International*, Jan. 1997, pp. 59–66.

[19] White, D., "Electromagnetic Compatibility: What Is It? Why Is It Needed?" *Instrument & Control Systems*, Jan. 1995, pp. 65–74.

[20] *SAE Handbook*, Vol. 2, *Parts and Components*, Warrendale, PA, 1995.

[21] Mladenovic, D., R. Verma, and R. Frank, "EMC Considerations for Automotive Sensors," SAE Internat'l Congress & Expo, SAE 970850, Detroit, Feb. 24–27, 1997.

12

Standards for Smart Sensing

The laws of the Round Table were laid down and every knight of the fellowship swore to keep the laws.
—John Steinbeck, *The Acts of King Arthur and His Noble Knights*

12.1 Introduction

Recent industry efforts have focused on the definition, functionality, and communication standards for smart sensors. The goal is interoperability in a wide range of applications. The smart sensor, with appropriate local decision-making capability, can act as a standalone sensor, communicate in a peer-to-peer relationship to other sensors and actuators, or act as an intelligent node in a network. Interoperability will allow the implementation of various interfaces across different networks for network independent operation. Such effort should expedite the development and transition to networked smart transducers as well as development and commercialization efforts for smart sensors and smart actuators.

12.2 Setting the Standards for Smart Sensors and Systems

In 1993, the IEEE and the National Institute for Standards and Testing (NIST) initiated an activity that has led to the development of two accepted and two proposed standards [1]:

- IEEE 1451.1 Network Capable Application Processor Information Model (approved by IEEE as full-use standard);

- IEEE 1451.2 Transducer to Microprocessor Communication Protocol and Transducer Electronic Data Sheet (TEDS) Formats (approved by IEEE as full-use standard);

- IEEE P1451.3 Digital Communication and Transducer Electronic Data Sheet (TEDS) Formats for Distributed Multidrop Systems;

- IEEE P1451.4 Mixed-Mode Communication Protocols and Transducer Electronic Data Sheet (TEDS) Formats.

As noted, IEEE 1451.1 and 1451.2 have been balloted and accepted as full-use standards.

A number of new terms and acronyms were developed to deal with the elements of smart sensors in an interoperable network. The TEDS is a machine-readable specification of the characteristics of the transducer interface [2]. A smart transducer interface module (STIM) includes the TEDS and the supporting electronics including the transducer on the transducer side of the hardware interface to the network-capable application processor (NCAP). The NCAP is a device that supports a network interface, application functionality, and general access to the physical world via one or more transducers. A transducer-independent interface (TII) is a 10-wire digital communication interface that allows an NCAP or host to obtain sensor readings or actuator actions as well as request TEDS data [1].

As shown in Figure 12.1 [1], the four standards tie together the elements linking smart sensors (the STIMS), bus-interface modules, and mixed-mode transducers to the network through the NCAP. The standards represent an extensive effort to provide sufficient detail to achieve interoperability while allowing flexibility for manufacturers of components, subsystems, and systems.

Table 12.1 compares the 1451 family members [3]. The major common element is the TEDS or the ability to support a TEDS. The hardware interface, the distance between elements, the signal converter, and the measurement level differentiate the four family members. The approved IEEE 1451.1 and 1451.2 standards are thorough documents that are hundreds of pages long. The remaining sections in this chapter provide more detail on the various family members, but interested readers are directed toward the actual documents and the references.

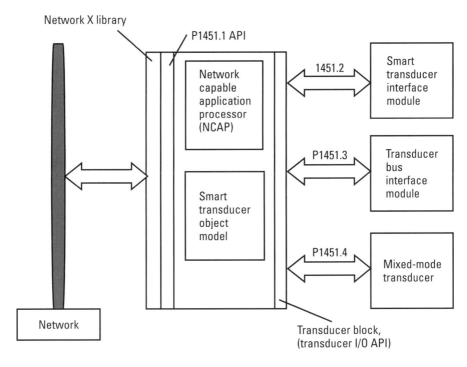

Figure 12.1 IEEE 1451 working relationship. (*After:* [1].)

Table 12.1
Summary of IEEE P1451 Family Members (*After:* [3])

Member	HW Interface	TEDS	Distance	Signal Converter	Measurement Level
1451.1	N/A	Support TEDS	N/A	N/A	Supports measurements
1451.2	10-wire digital	Yes	Short point-to-point	Yes	Raw digital engineering units
P1451.3	4-wire digital	Yes	Medium multi-drop	Yes	Raw digital engineering units
P1451.4	2- to 4-wire analog/digital	Yes	Medium point-to-point	No	Analog

12.3   IEEE 1451.1

The IEEE 1451.1 specification provides a simple, complete object model for building smart sensor and actuator-based systems [2]. In general, an object is a collection of data and operations. An object model is a definition of data structures and operations organized in a formal specification. For IEEE 1451.1, a smart transducer object model includes an interface to a transducer object model and to a transducer bus.

The IEEE 1451.1 standard models a smart transducer as a software PC with a backplane and plug-in cards [4]. The standard is intended for a manufacturer or integrator that provides additional functionality in the form of onboard intelligence that has to be supported on more than one underlying network. Figure 12.2 shows the building blocks and software components of parameters, events, actions, files, and timers.

12.3.1   Network-Capable Application Processor

The NCAP typically consists of a processor with an embedded operating system and timing capability. NCAPs can range from simple elements with more extensive support from STIMs with many channels to more complex designs with multiple ports. Figure 12.3 illustrates the top-level relations between block objects, software processes, and an NCAP [2]. Within an NCAP there can be several software processes in the function block, base transducer block, and

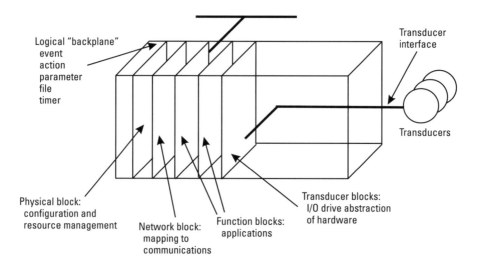

Figure 12.2   Object model components for IEEE 1451.1. (*After:* [4].)

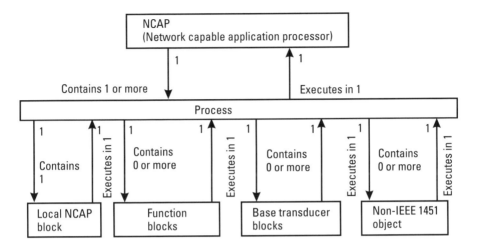

Figure 12.3 Top-level object relationships. (*After:* [2].)

non-IEEE 451 objects but only one object NCAP block. Object-identifying properties are the object's

- Class ID;
- Class name;
- Object ID;
- Object tag;
- Object name;
- Object dispatch address.

The IEEE 1451.1 standard provides both physical and logical specifications for the smart transducer object model [2]. In Figure 12.4, the solid lines represent the physical components of the system, and the dotted lines provide the logical view. The sensors and the actuators form a transducer that is connected over an interface to a microprocessor or microcontroller. The network protocol logical interface and transducer logical interface are defined in 1451.1. IEEE 1451.2 defines the (hardware) interface between transducers and the NCAP. The interfaces are optional, and either 1451.1 or 1451.2 can be implemented without the other. If support for interoperable transducers is not required, only 1451.1 could be used. If networking is not required, IEEE 1451.2 would be sufficient.

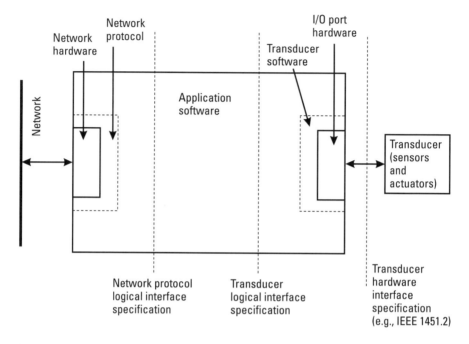

Figure 12.4 Networked smart transducer model. (*After:* [2].)

12.3.2 Network Communication Models

The IEEE 1451.1 standard provides two models for network communication between objects. The point-to-point client/server model is tightly coupled for one-to-one communications. The publish-subscribe model is loosely coupled for one-to-many and many-to-many communications. Network software suppliers are expected to provide code libraries that contain routines for the calls between the IEEE 1451.1 communication operations and the network [2].

Figure 12.5 shows the client-server used in IEEE 1451.1 [2]. The model is supported by two complementary application-level operations:

- Execute on client-side client-port objects;
- Perform on all network-visible server-side objects.

The execute and perform operations work together to provide a remote-object-operation-invocation-style messaging service [2].

The publish-subscribe model in IEEE 1451.1 is shown in Figure 12.6 [2]. The model provides a means for loosely coupled communications between

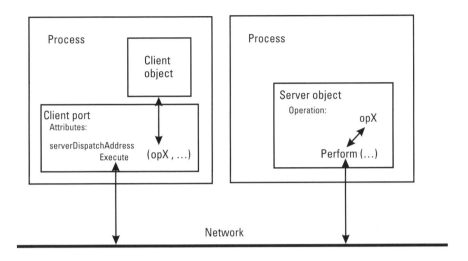

Figure 12.5   Client/server communications components. (*After:* [2].)

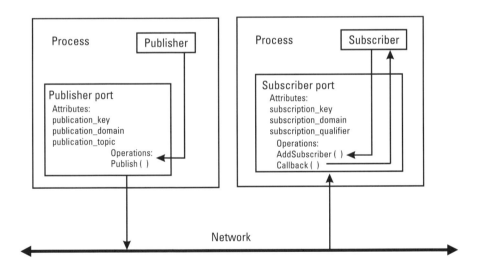

Figure 12.6   Publish-subscribe communications components. (*After:* [2].)

objects where the sending, or publishing, object does not need to be aware of the receiving, or subscriber, objects. This model is supported by two operations:

- Publish on publisher port objects;
- AddSubscriber and an associated callback operation on subscriber port objects.

The publisher and subscriber use a combination of domains, keys, and topics/qualifiers (defined in the standard) to allow only the publications of interest to a subscriber object to be selected from publications received by a subscriber port [2].

12.3.3 The IEEE 1451.1 Example

An example of how the NCAPs for sensors and actuators could be established for a wastewater treatment system is provided in the appendix of 1451.1. That example provides a rigorous implementation of the standard. Identifying the NCAPs is one of the initial steps. The system diagram in Figure 12.7 shows a

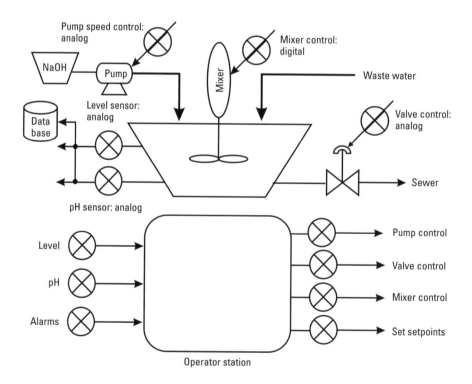

Figure 12.7 Wastewater treatment system example. (*After:* [2].)

PID control system (see Section 7.1.3) that takes periodic pH sensor readings to control the pump speed based on the measurements and pH set points [2].

The mixer control is either on or off, based on the pump speed being above or below a given set point. The measurements are logged to a database, used in the automated control system and displayed for an operator to override the automated system. Figure 12.8 shows the NCAP hardware required to implement the system with IEEE 1451 functionality.

The functionality of the wastewater system is partitioned among three NCAPs [2]. The level control and pH control are implemented on simple NCAPs and implemented separately to meet safety requirements. The operator system is implemented with a PC NCAP as the host for the system's operator-machine interface and data management system. Developing the software functionality for the system provides a rather complete use of 1451.1. Interested readers should refer to the standard for the full details of the example.

12.4   IEEE 1451.2

IEEE 1451.2 defines an interface, both the hardware and software blocks, for a networked transducer that does not depend on a specific control network [5].

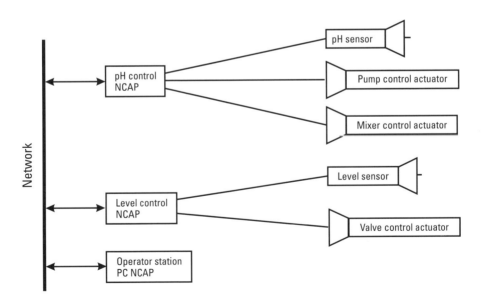

Figure 12.8 Hardware components of wastewater treatment system. (*After:* [2].)

The standard introduces the concept of a STIM, defines a TEDS, which is an integral part of the STIM, and the TII, the physical interface between the STIM and the NCAP [6]. Figure 12.9 shows the relationship between the elements defined in IEEE 1451.2 and the network [5]. Note that a single sensor or actuator or many channels of transducers may exist in a single STIM. This section explains key aspects of the STIM, TEDS, TII, and the built-in tools for enabling smarter systems.

12.4.1 STIM

The STIM in Figure 12.9 can contain from 1 to 255 transducers of various types [5]. A STIM is controlled by an NCAP module through a dedicated digital interface. A STIM meets the requirements of 1451.2 if it:

- Supports the required performance;
- Contains a properly formatted TEDS;
- Has a physical interface that implements the lines, protocol, and timing of the TII.

IEEE 1451.2 defines four types of sensors, one actuator, and an event detector [5]. The four sensor types all read some variable, convert the data from analog to digital form, and make the data available. There are six types of transducer:

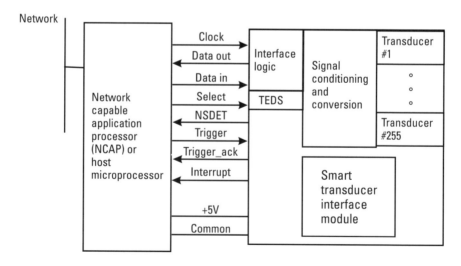

Figure 12.9 Hardware partitioning proposed by IEEE 1451.2. (*After:* [5].)

- Sensor;
- Buffered sensor;
- Data sequence sensor;
- Buffered data sequence sensor;
- Actuator;
- Event sequence sensor.

Figure 12.10 illustrates the response of a STIM to a trigger [6]. The triggering function provides a means for an NCAP to send the STIM a command for an action to take place (the trigger signal) and for the STIM to indicate the time when the action occurred (trigger acknowledgment). That general flow is elaborated in the standard to show the sensor and actuator activity that occurs concurrently in the quiescent and triggered states. Other possible triggering options include:

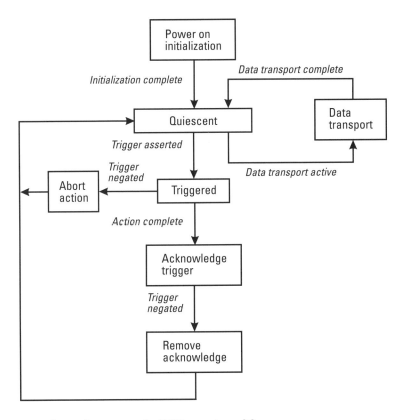

Figure 12.10 General response of a STIM to a trigger [6].

- Triggering buffered sensors;
- Triggering data sequence sensors;
- Triggering buffered data sequence sensor;
- Triggering event sequence sensors.

When the sensor STIM reacts to an NCAP trigger, the transducer begins reading [7]. When the reading is complete, the STIM makes a trigger acknowledge and the NCAP reads the value(s). The actuator STIM reacts to an NCAP trigger by writing the value(s) and initiating transducer actuation. When complete, the STIM asserts a trigger acknowledge.

Triggering the buffered sensor results in the same response as the standard sensor except the value is returned from the previous trigger. That means it is an immediate response, but the time of the reading is uncertain [7].

A data sequence sensor samples at its choice of time, typically synchronized with the physical world. After the trigger, the sensor waits until the next sample time and then returns an ACK signal and the results [7].

Triggering the buffered data sequence sensor is similar to triggering the data sequence sensor, except that it makes the data in the holding buffer available. The trigger acknowledgment is coincident with the availability of the previously acquired data set. If another trigger occurs after the holding buffer is read but before the acquisition of the next sample of data is complete, the acknowledge signal is not returned until the data acquisition process is complete [6].

An event sequence sensor is the same as the data sequence sensor, but it does not return any data. For this sensor, the time of the event is the relevant information [6].

12.4.2 Transducer Electronic Data Sheet

The TEDS identifies essentially everything you wanted to know or may want to know about the transducer, including:

- Manufacturer, model number, revision code, serial number, device type, and date code for transducers;
- When the unit was calibrated, the variable, type, and limits of use;
- Calibration constants;
- Signal conversion data model, model length, and number of significant bits;

- Channel timing read/write setup time, sampling period, warm-up time, and update time;
- Power supply requirements (voltage and current);
- Overhead: the TEDS length and number of channels.

Eight different TEDS memories are defined in 1451.2. The types of TEDS are either for use by the NCAP (machine readable) or for operators (human readable). There are two mandatory and four optional TEDS classifications. Table 12.2 lists the types of TEDS that are defined in 1451.2 [5].

Figure 12.11 shows the addressable sections of the TEDS [8]. The mandatory sections are shown with solid lines, the optional sections with dotted lines. The standard was defined to provide a growth path for future applications.

Table 12.3 shows an example of one of the TEDS in 1451.2—the data structure in the mandatory channel identification data block [6]. The 17 fields provide manufacturer- and sensor-specific information. The function of the channel identification TEDS is to make available at the interface all the information to identify the channel being addressed. The channel identification TEDS bytes are constant and read only [6].

12.4.3 TII

The standard digital interface for IEEE 1451.2, TII, is a clocked serial interface similar to an SPI (see Section 6.9.1). The 10 electrical connections are defined

Table 12.2
Types of Transducer Electronic Data Sheets (TEDS) [5]

Type	Readable By	IEEE 1451.2 Mandate
Meta TEDS	Machine	Mandatory
Channel TEDS	Machine	Mandatory
Calibration TEDS	Machine	Optional
Generic-extension TEDS	Machine	Optional
Meta-ID TEDS	Human	Optional
Channel-ID TEDS	Human	Optional
Calibration-ID TEDS	Human	Optional
End-user application-specific TEDS	Human	Optional

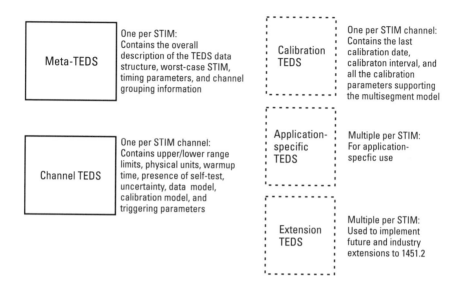

Figure 12.11 General layout of TEDS [8].

as shown in Table 12.4 [9]. The lines are defined more thoroughly in the standard, and the power supply aspects are discussed further in Section 12.4.5. The standard does not define a standard connector but leaves that as a flexibility item to the discretion of the user.

12.4.4 Calibration/Correction Engine

IEEE 1451.2 describes a correct algorithm that allows users to convert ADC output into engineering units for sensors and engineering units into input to an ADC for actuators [10]. Lookup tables, linear conversions, and multiple-input polynomial surface calibrations are among the possible uses of those data for sensors. The calibration TEDS contains the capability to perform:

- First-order conversion with a single segment straight line;
- Segmented first-order conversion for nonlinear sensors where higher accuracy is required;
- Level detector for simple two-state detection;
- Level detector with hysteresis;
- Higher order conversion with a single segment;
- Multiple-input conversion for transducers that need to achieve the highest possible accuracy.

Table 12.3
Data Structure of Channel Identification TEDS Data Block [6]

Field #	Description	Type	# Bytes
	Data structure related information data sub-block		
1	Channel identification TEDs length	U32L	4
2	Number of languages = L	U8C	1
3	String language code list Field 4 through 16 are repeated L times, once for each supported language	Array of U8E	L
	Identification related information data sub-block		
4	Language sub-block length	U16L	2
5	String specification	Lang	3
6	Manufacturer's identification length	U8L	1
7	Manufacturer's identification	String	≤255
8	Model number length	U8L	1
9	Model number	String	≤255
10	Version code length	U8L	1
11	Version code	String	≤255
12	Serial number length	U8L	1
13	Serial number	String	≤255
14	Channel description length	U16L	2
15	Channel description	String	≤65 535
	Data integrity information data sub-block		
16	Checksum for language sub-block	U16C	2
	Data integrity information data sub-block		
17	Checksum for channel identification TEDS	U16C	2

Figure 12.12 is an example of where the correction algorithm can be run [10]. Figure 12.12(a) shows the calibration TEDS stored in the STIM. In this case, the host computer copies the calibration TEDS from the STIM and performs the conversion from raw data to engineering units. That can be cost effective in small systems, but larger systems with many transducers can expend much of the CPU's time in the correction process. Distributed systems can have the conversion performed in one computer that provides corrected data to other processors or in each computer in the system. If a single processor

Table 12.4
Signal and Control Lines for the TII [9]

Line	Logic	Driven By	Function
DIN	Positive logic	NCAP	Transports address and data from NCAP to STIM
DOUT	Positive logic	STIM	Transports data from STIM to NCAP
DCLK	Positive logic	NCAP	Positive-going edge latches data on DIN and DOUT
NIOE	Active low	NCAP	Signals that the data transport is active and delimits data transport framing
NTRIG	Negative logic	NCAP	Performs triggering function
NACK	Negative logic	STIM	Serves two functions: trigger acknowledge and data transport acknowledge
NINT	Negative logic	STIM	Used by the STIM to request service from the NCAP
NSDET	Active low	STIM	Used by the NCAP to detect the presence of a STIM
POWER	N/A	NCAP	Nominal 5-V power supply
COMMON	N/A	NCAP	Signals common ground

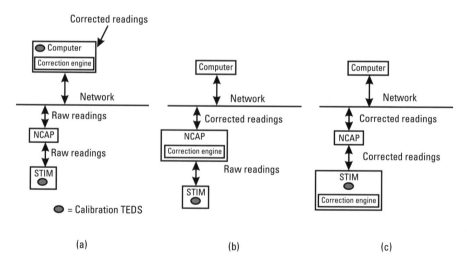

Figure 12.12 IEEE 1451.2 correction engine in the (a) computer, (b) NCAP, and (c) STIM. (*After:* [3, 10].)

performs the correction, the other processors must be programmed to get data only from the host and not directly from the transducer or the NCAP. If the conversion is performed in multiple places, the copy of the calibration TEDS must be changed each time a transducer is changed.

Figure 12.12(b) shows the conversion performed in the NCAP. The calibration TEDS is copied from the STIM into the NCAP, where the conversion to engineering units is performed. That simplifies the management of the source of data and eliminates multiple copies of the data. The NCAP processor will cost more to provide this capability.

Figure 12.12(c) shows the STIM running the correction process. This technique is useful in systems with a large array of transducers or with groups of very high speed transducers. For wide acceptance, ICs must be developed that provide this capability.

12.4.5 Sourcing Power to STIMs

Voltage on the power line at the NCAP is specified at 5V ±0.20V dc with respect to the common. Power for the STIM interface control circuitry must be provided only through the primary communications interface. Power normally is supplied to the STIM as shown in Figure 12.13(a) [6]. An alternative source of power for the STIM for applications where the sensor or actuator circuitry exceeds the current or voltage levels that the NCAP can provide is shown in Figure 12.13(b).

12.4.6 Representing Physical Units in the TEDS

IEEE 1451.2 has a simple and easily stored method of identifying physical units for use with smart transducers, allowing them to provide output in terms any user can understand [11]. This capability enhances sensor plug-and-play without writing special software. The approach uses a standard set of units to determine the calibration constants. For manufacturers, that means they can always perform final tests and calibrations the same way. For users, a standard technique exists for a variety of system functions by understanding the methodology used in IEEE 1451.2. The standard represents the International System of Units (SI) units, as listed in Table 12.5.

For an example of how to express a measurement of distance, consider (12.1) [11], which shows how a measurement in meters could be written. To provide this to a computer, only the exponential values are used, that is, 1, 0, 0, 0, 0, 0, 0, 0.

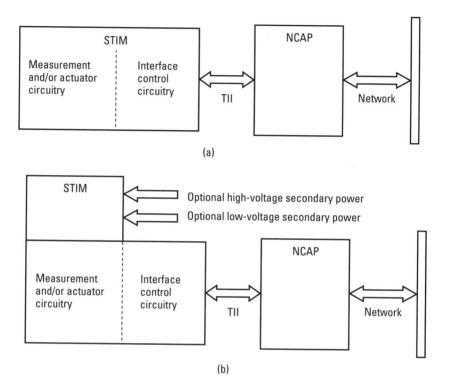

Figure 12.13 NCAP to STIM power: (a) single source from NCAP, and (b) NCAP with optional secondary power sources [6].

Table 12.5
SI Base Units in IEEE 1451.2 [10]

Base Quantity	Name	Symbol
Length	meter	m
Mass	kilogram	kg
Time	second	s
Electric current	ampere	A
Thermodynamic temperature	kelvin	K
Amount of substance	mole	mol
Luminous intensity	candela	cd

$$m^1 \cdot kg^0 \cdot s^0 \cdot A^0 \cdot K^0 \cdot mol^0 \cdot cd^0 \qquad (12.1)$$

The standard provides two derived units, radians (rad) and steradians (sr) to the seven base SI units and addresses units that are given in meters per meter, logarithm of quantities, and logarithms of dimensionless numbers, as well as hardness and digital data through an enumeration field (enum). Exponents are encoded using an unsigned byte integer. The exponent and its sign are multiplied by 2, and 128 is added to achieve an encoded byte integer with a resolution of $\frac{1}{2}$. Using this technique, exponents between −64 to +63 can be expressed.

As an example of how measurement units are handled in IEEE 1451.2, consider a pressure measurement that is expressed in pascals, that is, kilogram per meter per seconds squared. Table 12.6 shows the field units for this example. The enumeration 0 means that the units are the product of the base units. Examples of how distance, area, resistance, noise spectral density, mass fraction, strain, power quality, counts, and switch positions are shown in one reference [11].

12.5 IEEE P1451.3

IEEE P1451.3 defines a digital interface for connecting multiple, physically separated sensors [12]. This is one of two mixed-mode interfaces that allows digital information to be stored with the transducer and transmitted over analog data wires. The multidrop transducer bus standard is a minibus implementation small enough and cheap enough to integrate into a transducer. The amount of overhead for the proposed 1451.3 is considerably less than existing fieldbuses, which use up to 32 bits for node addressing [1]

Table 12.6
IEEE 1451.2 Pressure Measurement in Pascals [10]

Pressure (pascals = m − 1·kg · s − 2)									
	enum	rad	sr	m	kg	s	A	K	mol	cd
Exponent	0	0	0	−1	1	−2	0	0	0	0
Decimal		128	128	126	130	124	128	128	128	128

Figure 12.14 illustrates the interface for a minibus for multiple modules that are physically separated. The interface will supports TEDS, as well as channel identification, hot swapping, and time-synchronization protocols. The transducer bus interface module (TBIM) will consist of one to N transducers, a P1451.3-defined TEDS, and interface logic to control and transfer data across the minibus. Most likely, the TEDS will contain a meta TEDS, a meta-ID TEDS, a channel TEDS, a channel-ID TEDS, a calibration TEDS, and a calibration-ID TEDS but with different names to avoid confusion with those defined in 1451.2. The transducer bus controller (TBC) will manage the minibus and handle the setup and data transfer [12].

12.6 IEEE P1451.4

The amount of existing sensor and networking technologies and the cost of transitioning to the digital network standards defined in 1451.1, 1451.2, and P1451.3 are issues that the proponents of the standard are addressing. Analog sensors, with their existing wiring and the requirement for wide bandwidth analog measurements are being addressed by P1451.4. The proposed IEEE 1451.4 will establish a standard that allows those analog-output, mixed-mode transducers to communicate digital information with a high-level IEEE 1451

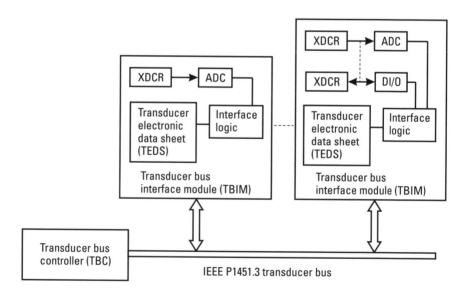

Figure 12.14 IEEE 1451.3 interface specification proposal. (*After:* [12].)

object [13]. To fit into the digital network defined by other 1451 standards, the bidirectional digital communication of self-identification, test, and programmable signal conditioning is being defined with consideration for simplicity and low cost. P1451.4 will offer compatibility with legacy systems and provide a transition path to 1451 [13]. Both sensors and actuators are supported by P1451.4, and yet the interface will be invisible from the network's perspective.

Figure 12.15 shows the TEDS and high-level P1451 object for an analog transducer [13]. For many P1451.4 applications, it is not practical to include the network interface (the NCAP) with the transducer, because of size limitations or harsh operating environment considerations. However, the transducer TEDS must contain enough information to allow the higher level P1451 object to fill in any gaps.

Figure 12.16 is an example of an IEEE P1451.4 implementation. The NCAP and TEDS characteristics are defined in 1451.1 and 1451.2 and refined as necessary for a mixed-mode network in P1451.4. Minimizing the amount of nonvolatile memory is one goal of the standard's developers. The 1451.4 TEDS includes the categories and parameters listed in Table 12.7 [13].

12.7 Extending the System to the Network

The IEEE 1451.2 standard instigated industry design activity based on its anticipated approval [14]. Its subsequent approval and the addition of IEEE

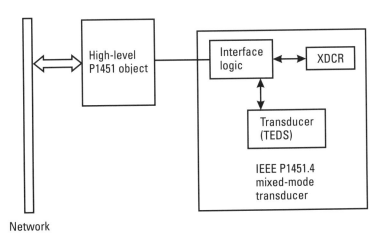

Figure 12.15 IEEE P1451.4 interface. (*After:* [13].)

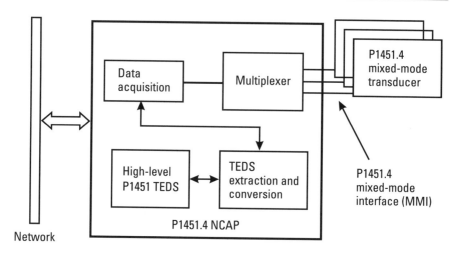

Figure 12.16   IEEE P1451.4 NCAP. (*After:* [13].)

Table 12.7
Categories and Identification in Proposed 1451.4 TEDS (*After:* [12])

Category	Parameters
1. Identification	Manufacturer name Model number Series number Revision number Date code
2. Device	Sensor type Sensitivity Bandwidth Units Accuracy
3. Calibration	Last calibrated date Correlation engine coefficients
4. Application	Channel identification Channel grouping Sensor location and orientation

1451.1 should expand industry support. As shown in Figure 12.17, the network independence provided by 1451 allows manufacturers to connect smart sensors in different systems to various networks.

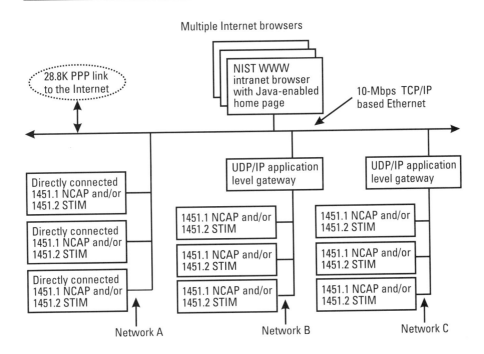

Multiple Internet browsers

28.8K PPP link
to the Internet

NIST WWW
intranet browser
with Java-enabled
home page

10-Mbps TCP/IP
based Ethernet

UDP/IP application
level gateway

UDP/IP application
level gateway

Directly connected
1451.1 NCAP and/or
1451.2 STIM

Directly connected
1451.1 NCAP and/or
1451.2 STIM

Directly connected
1451.1 NCAP and/or
1451.2 STIM

1451.1 NCAP and/or
1451.2 STIM

1451.1 NCAP and/or
1451.2 STIM

1451.1 NCAP and/or
1451.2 STIM

1451.1 NCAP and/or
1451.2 STIM

1451.1 NCAP and/or
1451.2 STIM

1451.1 NCAP and/or
1451.2 STIM

Network A

Network B

Network C

Figure 12.17 Networked smart sensors [15].

12.8 Summary

Standards have played a major role in many industries to accelerate and expand acceptance. The accepted and proposed IEEE 1451 standards described in this chapter have been or are being developed to work together or separately, depending on system requirements. These standards should provide transducer-to-network interoperability and extended benefits to existing sensor manufacturers and network users. The potential to use the Internet in control systems will change the way networks are designed by engineers and used in organizations. Chapter 13 expands the future potential and the visions that are developing for networked smart sensors.

References

[1] Lee, K., "An Overview of the IEEE 1451—A Family of Proposed Smart Transducer Interface Standards," *Proc. Sensors Expo*, San Jose, CA, May 19–21, 1998, pp. 159–167.

[2]   "Smart Transducer Interface for Sensors and Actuators-Network Capable App. Processor,"
      IEEE 1451.1 Standard, http://standards.ieee.org/catalog/drafts.html, Piscataway, NJ, Apr.
      1999.

[3]   Woods, S., "IEEE P1451" Presentation, *Sensors Expo*, San Jose, CA, May 20, 1998.

[4]   Warrior, J., "The IEEE P1451.1 Object Model Network Independent Interfaces for Sen-
      sors and Actuators," *Proc. Sensors Expo*, Boston, May 13–15, 1997, pp. 1–14.

[5]   Eccles, L. H., "A Brief Description of IEEE P1451.2," *Proc. Sensors Expo*, San Jose, CA,
      May 19–21, 1998, pp. 169–180.

[6]   "IEEE Standard for a Smart Transducer Interface for Sensors and Actuators—Transducer
      to Microprocessor Communication Protocols and Transducer Electronic Data Sheet
      (TEDS) Formats," IEEE 1451.2 Standard, http://standards.ieee.org/catalog/olis/im.html,
      Piscataway, NJ.

[7]   Hamilton, B., S. Woods, and R. Johnson, "Designing IEEE 1451.2 Smart Transducer
      Interfaces," Presentation handout at Sensors Expo, Detroit, Oct. 22, 1997.

[8]   Woods, S., "The IEEE-P1451.2 Draft Standard for Smart Transducer Interface Mod-
      ules," *Proc. Sensors Expo*, Boston, May 13–15, 1997, pp. 5–14.

[9]   Cantrell, T. "Car 1451, Where Are You?" *Circuit Cellar Ink* (http://www.
      edtn.com/embapp/emba041.htm), Feb. 1999.

[10] Eccles, L. H., "IEEE-1451.2 Engineering Units Conversion Algorithm," *Sensors*, May
      1999, pp. 107–112.

[11] Eccles, L., "The Physical Representation of Physical Units in IEEE 1451.2," *Sensors*, Apr.
      1999, pp. 28–35.

[12] Malchodi, L. A., "Status of IEEE P1451.3 (Distributed Multidrop Systems)," *Proc. Sensors
      Expo*, San Jose, CA, May 19–21, 1998, pp. 181–184.

[13] Chen, S. C., "IEEE-P1451.4 A Smart Transducer Interface for Sensors and Actua-
      tors—Mixed-Mode Communication Protocols and Transducer Electronic Data Sheet
      (TEDS) Formats," *Proc. Sensors Expo*, San Jose, CA, May 19–21, 1998, pp. 185–191.

[14] Travis, B., "Sensors Smarten Up," *EDN*, Mar. 4 1999, pp. 76–86.

[15] Schneeman, R., and K. Lee, "Multi-Network Access to IEEE P1451 Smart Sensor Infor-
      mation Using World Wide Web Technology," *Proc. Sensors Expo*, Boston, May 13–15,
      1997, pp. 15–34.

13

The Implications of Smart Sensor Standards

And their marvelous instruments still continued to function, watching over
the experiments started so many years ago.
— Arthur C. Clarke, *3001: The Final Odyssey*

13.1 Introduction

Industry standards for smart sensors, including the IEEE 1451 family and others that have been initiated for control applications, should accomplish what their architects envision: reducing the barriers to acceptance and therefore accelerating the development and use of new smart sensors in existing and advanced systems. Some of the capabilities that the IEEE 1451 standards will enable have already been demonstrated. Other possibilities are being proposed and developed by companies with a vision for the future. This chapter discusses how plug-and-play makes sensors smarter and easier to deal with in the real world and the advanced communication and control capabilities that smart sensors can bring to control systems, the Ethernet and the Internet.

13.2 Sensor Plug-and-Play

Using a computer technology and an object-oriented technology approach to sensing is behind several standard approaches. In addition to the IEEE 1451 standard, Microsoft has developed object linking and embedding (OLE) for supporting PCs in embedded control applications. The OLE for Process Control (OPC) Foundation is driving commonality based on OLE-based standard objects, methods, and properties for servers of real-time information in distributed control systems, programmable logic controllers, smart field devices, and analyzers [1]. Object technology answers the demands for open systems in control system design. Examples of objects include a valve, loop, trend log, field device, pressure sensor, and I/O board. Objects can model real-world data, or the data contained within the object may be real time. Figure 13.1 shows how object models could control operations and events in a PID control system.

The use of Java program language for embedded applications has interesting implications for smart sensors [2]. A Java automation application programming interface (API) is being developed by Sun Microsystems and a group of process control and manufacturing systems vendors. Java computes platform independent and will run on any Java-enabled machine. Applications are decoupled from platforms and can be distributed across networks on servers with full or partial applications delivered to clients. In such an environment, only one copy of the application is maintained on the server. The user always has the current version and the application goes away when the user's task is completed [1].

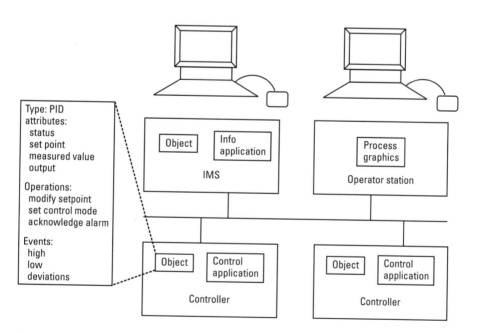

Figure 13.1 Object model for control. (*After:* [1].)

Cost is a key enabler for many new systems. As software costs and development times continue to rise, many companies in the embedded-systems world are viewing Java as a program language that will let them write code once and move easily from one MCU to another [2]. However, Java was developed for desktop computing and has functionality that exceeds the requirements of embedded control. Subsets of Java could prove to be the answer for more cost-control and memory-sensitive applications.

The PC serial buses universal serial bus (USB) and IEEE 1394 (also known as Firewire) provide plug-and-play and are expected to play a prominent role in data acquisition (DA) [3]. Table 13.1 [3] compares those two plug-and-play standards. As many as 127 devices can be connected to one USB

Table 13.1

USB Compared to Firewire for Data Acquisition (*After:* [3])

Bus	Pros	Cons	Theoretical Throughput	Actual Throughput	Distance From PC
ISA bus	Large variety of boards, extensive software support, easy to write drivers, cost-effective boards	Slow data transfer, few plug-and-play boards	2–3 MBps	500–750 Ksps	Internal
PCI bus	Plug-and-play, high-speed bus transfers	Limited number of PCI slots in some computers	132 MBps	80–100 MBps	Internal
PC cards	Small size—portability	Limited channel count, no DMA support	Machine dependent	200–300 Ksps	Up to 30–50 ft
Enhanced parallel port	Outside the PC—noise immunity	Limited transfer rates	500 Ksps burst	100 Ksps continuous	5 m/cable
Universal serial bus	Outside the PC—noise immunity, true plug-and-play	Found only in new purchase, limited data transfer rates	12 Mbps	<100 Ksps	4.5 m/cable
FireWire (IEEE 1394)	Outside the PC—noise immunity, high-speed transfers, automatic configuration	Not yet installed on many PCs	400 Mbps	Not available	4.5 m/cable

port. When these PC-based standards are applied to sensing, the level of smart will increase.

13.3 Communicating Sensor Data Via Existing Wiring

One way to implement advance sensing quickly is to use existing standards with hardware and software developed specifically for sensing. Ethernet is established in the plant environment. Telephone lines also are being used for sensing. This section discusses both applications.

13.3.1 Ethernet

The existing Ethernet wiring in many plants allows the retrofitting of sensors to monitor systems for preventative maintenance and avoid costly downtime [4]. The capability of speeds up to 1 Gbps and the ability to interface directly to industrial and personal computers make the Ethernet a low-cost alternative to other LANs. Most available data acquisition systems can communicate with a 10Base-T Ethernet.

More sophisticated software is required for the real-time aspect of data acquisition. That is especially true for a network shared with general business users that could be disrupted [5]. Different applications such as data monitoring versus data logging place different requirements on the software. In data logging, the connection to the computer is broken after data collection configuration. In this case, the DA device must have a high level of intelligence to store the configurations and send data to a buffer without communication with external software. Data monitoring applications require less intelligence in the DA device and more in the software to perform tasks typically handled in real time.

13.3.2 Sensing by Modem

Companies are already manufacturing modem units specifically for sensing applications. One unit, the PhoneDucer™ from Elwood Corp., sends formatted sensor data over telephone lines [6]. Power for the modem and the sensors comes from the telephone line so no additional power supply or battery is required. The modem allows engineers to collect, monitor, and analyze data from a remote site using a PC running Windows and special DA software. Natural gas utilities and chemical manufacturing companies are among the companies that are taking advantage of this remote sensing approach.

13.4 Automated/Remote Sensing and the Web

The major features of IEEE 1451.1 and IEEE 1451.2 standards were demonstrated at Sensors Expo Boston in May 1997 [7]. The demonstration included participants from sensor and transducer manufacturers, control network vendors, system integrators, and users. The networked 1451 devices were accessible via a common 10-Mbps Ethernet backbone. Figure 13.2 shows the architecture concept used in the demonstration. The integration of multivendor networks and devices using the common set of interfaces of IEEE 1451 was demonstrated with Internet-based technologies, including World Wide Web (WWW) browser software, Java, HTML Web pages, and the Internet protocol (IP) suite.

Figure 13.3 illustrates a reference implementation of the IEEE 1451 standard [9]. The STIM in the figure contains a low-cost microcontroller programmed to support the 1451 digital interface. The microcontroller reads the TEDS and gathers sensor readings from the ADC. The dotted-line blocks perform both firmware and software functions. The firmware on the NCAP accesses the STIM through a 1451.2 driver. The correction engine corrects the

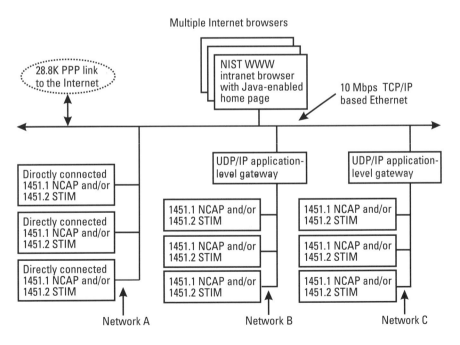

Figure 13.2 Conceptual layout of NIST 1451 demonstration. (*After:* [8].)

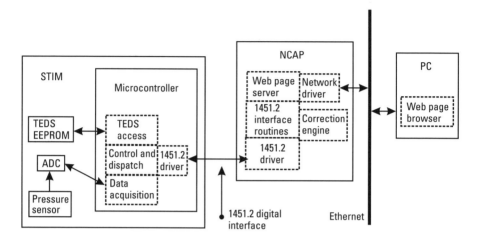

Figure 13.3 Reference implementation of IEEE 1451. (*After:* [9].)

raw sensor readings from the STIM. The NCAP's Web page is viewed by pointing a Web browser at the NCAP. The prototype supports a wide variety of analog sensors.

13.4.1 Wireless Protocol

Some remote measurements and wireless sensing applications may require more industry effort or benefits to make the standard attractive. For example, a single-chip wireless sensor that transmits sensor information in shipboard conditions does not—at this time—use IEEE 1451 [10]. As shown in Figure 13.4, the fully integrated wireless sensor has four different sections: (1) sensors and front-end signal processing, (2) digitization, control, and spreading-code generation, (3) RF transmitter, and (4) the antenna. The lower left corner is an area designed to be sensitive to a specific frequency of an IR signal. The chip contains a communication protocol that allows a TV-style remote to program the sample rate, analog gain, and other parameters. The design has reliably collected and transmitted data over three decks of a ship without incurring EMI problems. Section 13.6 discusses alternative standards.

13.4.2 Remote Diagnosis

Remote measurements are allowing physicians to analyze and treat patients in remote areas [11]. Applications for that approach include transmission from an overseas battlefield or work site, telemedicine for hospitals in rural areas, remote

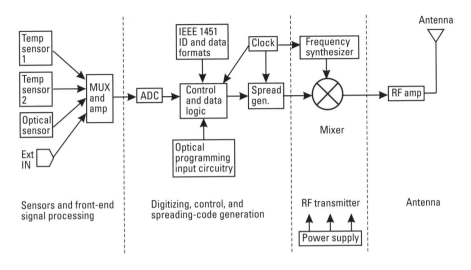

Figure 13.4 Intelligent wireless sensor. (*After:* [10].)

confirmation of diagnoses, and access to doctors in other regions of the world. The applications determine the requirements for the data transmission. Stethoscope measurements require high-quality sound captured down to 10 Hz. The system shown in Figure 13.5 allows a doctor in a remote location to hear what a nurse or medic hears when examining a patient. The remote doctor uses stethophones after the signal has been transmitted and processed by a codec.

The Internet is being used for more complex measurements that require data transmission. In the area of medical research, a PC and WWW browser software allow users to view magnetic resonance imaging (MRI) data from a remote 170-MHz proton imaging spectrometer. That means a number of doctors can be trained to perform modern diagnostic techniques that require expensive equipment without having to be located in a facility that has the equipment. The vision of interactive, real-time experiments over the WWW should increase the number of people doing advanced research and could drive discovery of new cures and therapy for patients everywhere.

13.5 Process Control Over the Internet

Transmission control protocol (TCP) and IP have been discussed as a means of monitoring but their role could be extended to the control systems [12]. Typically the same software to control a data acquisition system via a network can be

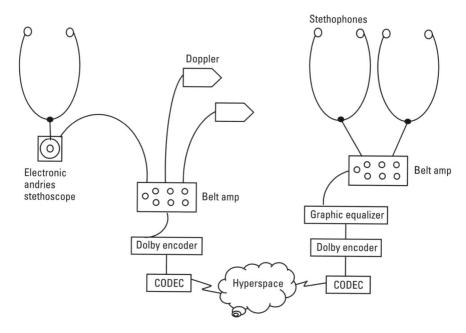

Figure 13.5 Telestethoscope, a stethoscope that allows a remote doctor to analyze a patient. (*After:* [11].)

used for the Internet as well [5]. Every sensor connected to a TCP/IP must have an IP address. The software must allow the proper configuration of the sensors.

Monitoring of remote locations via the Internet was discussed earlier in this chapter. Researchers also are working on telerobotics for remote control of machinery [13]. One of the key issues is a standard for interfacing modular portions of the system. A virtual environment has been created with an Internet extension that allows remote manipulation of the robotic system. The key communication element is the use of virtual reality modeling language (VRML), an extension of hypertext markup language (HTML). VRML allows developers to create programs that interact with the three-dimensional environment. Using VRML, generic sensors would be combined with robotic machinery to manipulate material and information that is passed to successive stages of the laboratory robotic. The extent or complexity of the robotic chain depends on solving a number of issues, including proper handling of queuing requests.

The effort for remote control in networks is already outside the laboratory and part of the industrial world. Figure 13.6 shows the general direction of future Internet software releases from one of the largest factory automation software suppliers [14]. The small ovals on each object represent scaled versions of available commercial software and provide the Internet-enabling properties

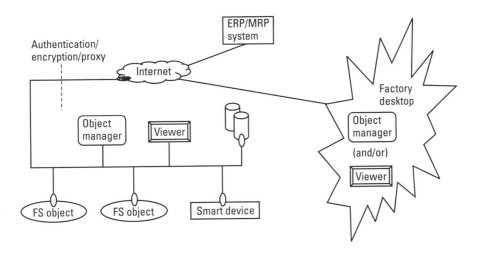

Figure 13.6 Internet software. (*After:* [14].)

of each object. The object manager and the viewer are visual objects for managing the Internet-enabling properties and viewing information in another of the company's software packages. The uniform and global network properties of the Internet will allow the object manager and the viewer to simultaneously engage both local and remote systems and other manufacturing systems. The file transfer Internet standards will allow offline editing and viewing.

Consumer control applications using the Internet could become common place in the future. For example, the refrigerator may someday have a universal product code (UPC) detector to monitor the expiration date of its contents [13]. Restocking orders could be placed to an online grocery for replacement, shipping, and billing. Robotic elements could fully squeeze oranges in the refrigerator for drinking or melons at the grocery to test for ripeness according to the buyer's preference.

13.6 Alternative Standards

Not all systems are covered by the standards or addressed in a manner that satisfies all applications. Critical and unique environments with design constraints outside the standards can require modifications of a standard and, in some cases, another standard. Two examples will be discussed, but they are just a sample of the instances when no matter how many "standards" are available another approach is deemed necessary.

13.6.1 Airplane Networks

Figure 13.7 shows the use of a variation of the IEEE 1451 standard for a jet airplane [15]. The primary difference is the analog output for troubleshooting. Each of a number of buses on the airplane would connect a number of sensors or actuators to a central DA and/or control system. The connection between the hub and the host computer would be less than 100m and actually may not be required. A single high-speed cable or fiber-optic link would provide the interface to the host computer, located remotely from the network. The development team analyzed other available buses and eliminated them for the following reasons:

- They were too slow to support desired data rates.
- They required four-wire cable.
- They had undesirable features.
- There was limited physical distance between points.

13.6.2 Automotive Safety Network

Cost-effective simple circuitry for networked control of airbags and other vehicle safety systems has led to the development of a distributed systems interface (DSI) [16]. The interface is offered as a freely licensable standard. Other

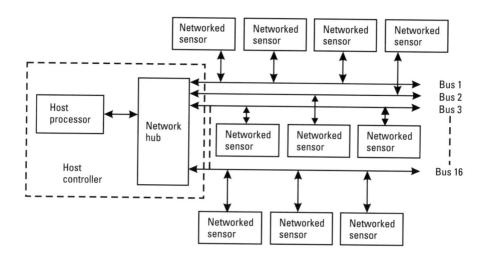

Figure 13.7 Boeing 777 adaptation of IEEE 1451.2. (*After:* [15].)

standards, such as CAN, or pending standards, such as the drive-by-wire bus TTP/C, were considered too expensive for the airbag application.

The DSI is not designed to be fault tolerant, but it has a high immunity to EMI. The signal protocol uses voltage mode signaling in one direction (from the master) and current mode signaling in the other direction (from the slave). However, the DSI does not fail without warning. During powerup, each node is analyzed, and the airbag can be added or removed, depending on the presence and the size of passenger, without completely reconfiguring the network. Configuration during powerup allows components from several different suppliers to be combined for the first time at vehicle assembly or to be replaced in service if they are found to be defective without special programming equipment [17].

Figure 13.8 is an example of the two-wire DSI network. The master node polls each sensor for information, calculates whether an airbag needs to be deployed, and then issues a command to the appropriate actuator (squib). The network that connects to other networks in the vehicle is called the niche area network (NAN).

Figure 13.9 shows a DSI network with the various components of an airbag system [17]. Although the network is a NAN, 23 components are identified in the figure. The ability of automakers to add safety components, such as side-impact airbags, kneebags, tubular restraints, pretensioners, and occupant sensors without redesigning the entire system provides flexibility and saves time and resources.

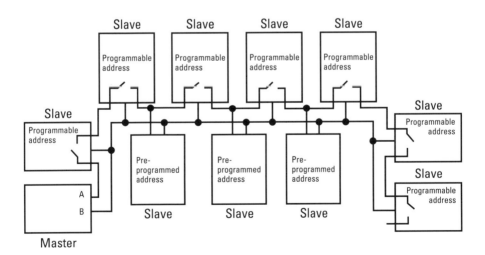

Figure 13.8 Example of a DSI. (*After:* [17].)

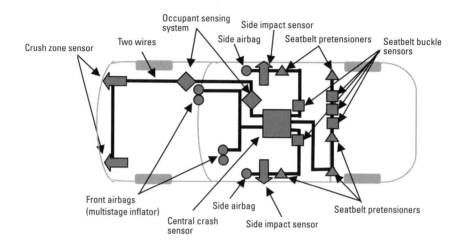

Figure 13.9 Automotive airbag system using the DSI standard. (Courtesy of Motorola and TRW.)

13.7 Summary

The initial stages of a smart architecture for sensors and networks are appearing in several different areas. The emphasis is being placed on interoperability and standardization of protocols, along with the flexibility offered by an array of components from building blocks to monolithic solutions. As a result, next-generation control systems can have a level of flexibility and functionality that will grow as the industry grows and expand as the applications require. However, no matter how many standards are available, applications may require variations or new alternatives, which eventually may become standards in their own right.

References

[1]   Lange, S. T., "What's All This About Object Technology?" *Control Engineering*, Aug. 1997, pp. 59–64.

[2]   Costlow, T., "Java Challenges Embedded-Systems World," *Electronic Engineering Times*, Oct. 28, 1996, pp. 79, 81.

[3]   Logan, L., "Are You on the Right Bus?" *Sensors*, Aug. 1997, pp. 46–56.

[4]   Daniel, R. A., "A Case for Communicating Sensor Data via Ethernet," *I&CS*, Oct. 1997, pp. 37–40.

[5] Lekas, S., "Making Ethernet Work: The Software Behind Successful Data Acquisition," *Sensors*, Feb. 1999, pp. 91–95.

[6] "Sensing by Modem," *Mechanical Engineering*, Sept. 1994, p. 34.

[7] Johnson, R. N., "Building Plug-and-Play Networked Smart Transducers," http://www.sensorsmag.com/ieee1097.htm

[8] Schneeman, R., and K. Lee, "Multi-Network Access to IEEE P1451 Smart Sensor Information Using World Wide Web Technology," *Proc. Sensors Expo Boston*, Boston, May 13–15, 1997, pp. 15–34.

[9] Woods, S. P., "The IEEE 1451.2 Draft Standard for Smart Transducer Interface Modules," Boston, May 13–15, 1997, pp. 5–14.

[10] Manges, W. M., G. O. Allgood, and S. F. Smith, "It's Time for Sensors to Go Wireless," *Sensors*, Apr. 1999, pp. 10–20.

[11] Machlis, S., "'Teledocs' Treat Patients Remotely," *Design News*, June 10, 1996, pp. 23–24.

[12] Winkler, R., "Clearing Up the Confusion on Ethernet," *Sensors*, Jan. 1999, pp. 28–35.

[13] "Remote Controlled Machinery, Processes Over the Web Closer Than You Think," *MIT Report*, Vol. XXV, No. 7, Sept. 1997, pp. 1–2.

[14] Carson, M., "Internet Strategy," http://www.wonderware.com/scout/index.htm

[15] Eccles, L., "A Smart Sensors Bus for Data Acquisition," *Sensors*, Mar. 1998, pp. 28–36.

[16] Edwards, C., "Motorola and TRW's Car Comm System 'Will Beat CAN and TTP/C,'" *Electronics Times*, Mar. 15, 1999, p. 10.

[17] DSI Bus Standard Release 1.0, http://www.mot-sps.com/automotive

14

The Next Phase of Sensing Systems

Once you have eliminated the impossible, whatever is left is possible.
—Ancestor of Mr. Spock, Chief Science Officer of the Starship *Enterprise*

14.1   Introduction

As Chapter 11 pointed out, the combination of several techniques can be used to develop the next generation of smart sensing systems. That has been possible for many years and is indeed the basis of many instruments, such as the Fabry-Perot interferometer, blood testing systems, and the camera-on-a-chip. What makes the future possibilities exciting is the scale (size) of the sensors made possible by the combination of micromachining and microelectronics. The size of semiconductor sensors is directly proportional to the cost of the sensor, and the cost is inversely proportional to the volume of applications. When cost barriers are broken, laboratory curiosities become a part of everyday life. Manifold absolute pressure sensors in cars, disposable blood pressure sensors for medical applications, and accelerometers for airbag systems have demonstrated the extent to which a cost-effective sensing technology can displace a previous technology. That is especially true when on-chip integration is possible.

Smart sensors can be standalone elements of simple products or distributed in complex networks such as a factory or a vehicle. The availability of standards such as the IEEE 1451 should expedite the development in networked sensors, especially the smarter versions.

In the area of computing, communications, and entertainment, the convergence of these technologies is creating entirely new products and markets.

Similarly, computing, communications, and other technologies will be combined with several or all of the topics that have been discussed in this book to create new control products. Those new products may require the combinations of several enabling technologies that are in the process of achieving manufacturing status. Aggressive semiconductor industry roadmaps, heavy investment in R&D for sensors, and the large number of participants in sensor technology and manufacturing promise a variety of competitive products, with the focus on the smart aspects of these sensors.

Many industry experts are actively trying to define the next "killer" applications for MEMS technology. In fact, several conferences on the commercialization of microsystems have been held for experts to share ideas in that area. As shown in Table 14.1, product evolution has encompassed decades to full commercialization phase for today's successes [1]. Smart sensors in gas, bio/chemical, and rate sensor products are among the EMS devices that are still in the evolution phase.

MEMS technology, which forms the basis of the smart sensors discussed in this book, has been called disruptive to industry according to a recent study [2]. "Disruptive" technology upsets the status quo in an industry. In the case of MEMS, the market pull is bringing process-technology and packaging companies into the regime of IC manufacturers. The study was funded by the National Science Foundation (NSF) and Semiconductor Equipment and

Table 14.1
Timetable of Evolution of MEMS/MST Products (*After:* [1])

Product	Discovery	Product Evolution	Cost Reduction/ Application Expansion	Full Commercialization
Pressure sensors	1954–1960	1960–1975	1975–1990	1990
Accelerometers	1974–1985	1985–1990	1990–1997	1997
Gas sensors	1986–1994	1994–1998	1998–2005	2005
Valves	1980–1988	1988–1996	1996–2004	2004
Nozzles	1972–1984	1984–1990	1990–2002	2002
Photonics/displays	1980–1986	1986–1996	1996–2004	2004
Bio/chemical sensors	1980–1994	1994–1999	1999–2002	2002
RF	1994–1998	1998–2001	2001–2005	2005
Rate sensors	1982–1990	1990–1998	1998–2002	2002
Microrelays	1982–1992	1992–1998	1998–2004	2004

Materials International (SEMI). Based on the disruptive implications of MEMS, established companies may not be the ones finding successful smart sensor applications.

14.2 Future Semiconductor Capabilities

Because micromachined sensors are based on semiconductor technology, a semiconductor industry roadmap could provide insight into future sensing technology. The Semiconductor Industry Association (SIA) periodically publishes technology predictions for future capabilities of leading-edge semiconductors. As shown in Table 14.2 [3], by 2012 the minimum feature size for the highest performance microprocessor circuits will be only 35 nm, only about one-sixth the size of a 1997 feature. The packaged cost will be reduced by more

Table 14.2
Projected Semiconductor Technology Capability

Characteristic	Year						
	1997	1999	2001	2003	2006	2009	2012
Minimum feature size (nm)	200	140	120	100	70	50	35
Logic transistors/cm^2 (packed)	3.7M	6.2M	10M	18M	39M	84M	180M
Cost/transistor @ volume (microcents)	910	525	305	175	75	34	15
Maximum number of wiring levels (logic)	6	6–7	7	7	7–8	8–9	9
Electrical defect density (d/m^2) 60%	1,940	1,710	1,510	1,355	1,120	940	775
Minimum mask count	22	22/24	23	24	24/26	26/28	28
Chip size (mm^2) ASIC	480	800	850	900	1,000	1,100	1,300
Power supply voltage (logic)	1.8–2.5	1.5–1.8	1.2–1.5	1.2–1.5	0.9–1.2	0.6–0.9	0.5–0.6
Number of chips I/OS (cost-performance)	800	975	1,195	1,460	1,970	2,655	3,585
Performance (MHz) (chip-board)	250	480	785	885	1,035	1,285	1,540

than 98%, dropping from 910 to 15 microcents per transistor for the third year of MPU production. The number of logic transistors will increase by over 48 times, from 3.7 million to 180 million. Unfortunately, there are no direct predictions about micromachining technology or sensors. However, the part that makes any sensor smart—the computational engine—is increasing at a phenomenal rate. At some point in the future, the supercomputer that performs today's laboratory calculations will be a portable computer with similar performance.

Sensing technology will play an important role in developing the next generation of semiconductor technology. Sandia National Laboratory has developed a silicon chip with up to 250 microsensors to monitor the mechanical, chemical, and thermal environments of integrated circuits [4]. The chip can be used during prototyping, manufacturing, or any time during the life of the chip to monitor critical parameters that could affect performance. The assembly test chip with its onboard polyimide and Al_2O_3 moisture sensors, piezoelectric strain gauges, electrostatic discharge and corrosion detectors, mobile ion detectors, and thermocouples promises to play an important role in developing semiconductor technology.

The scale of future semiconductors will require measuring and monitoring techniques that are well beyond today's capability, especially if they will be used on a day-to-day basis to monitor production processes and measure quality control. The use of contamination sensors for measuring and monitoring particulates and moisture is among the recommendations from the SIA roadmap [5]. Candidates for sensors include low-cost gas analysis sensors such as residual gas analysis, optical emission spectroscopy, and intracavity laser spectroscopy. The integration of such sensors into online equipment was recommended. Metal and total-oxidizable-carbon contamination measurements in liquids were also considered essential. The report frequently cites the need to monitor wafer contamination during processing as a key to achieving manufacturable product as the critical dimensions shrink. Undoubtedly, sensors that allow semiconductor manufacturers to achieve the predicted capability will be pursued and implemented as they are available. The Sematech (a consortium of semiconductor manufacturers) project to develop interoperable sensor and actuator standards confirms the industry's desire to communicate the information from those sensors in an intelligent or smart manner using a common protocol [6].

In-situ smart sensors for wafer metrology have a high payback potential for the semiconductor industry. According to one technologist, direct measurement of the wafer state, such as film thickness and film composition during processing, could save calibration setup, maintenance, and testing time [7].

Communication between the sensors and the tools is essential, an area in which the IEEE 1451 standards should help.

14.3 Future System Requirements

The unprecedented capability that semiconductor technology will provide raises other questions. How much technology will we really need? Who is going to apply that technology to smart sensors? The answers to those questions and several others that could be asked come from examples that are present today.

The extensive use of manifold absolute pressure sensors in vehicle emission control systems resulted from two factors: (1) government legislation forced a change in the way automotive engines were controlled, and (2) semiconductor-based sensors proved to be more cost effective and more reliable than the mechanical versions that were originally used to define the control system. Legislation and customer demand drove the development of technology.

In invasive blood pressure measurements, disposable semiconductor-based sensors initially proved to be lower cost than resterilizing and periodically recalibrating expensive mechanical units. As the replacement of mechanical units was proceeding at an as-predicted rate, the threat of the human immunodeficiency virus (HIV) accelerated the conversion process. Cost effectiveness and consumer demand drove this second sensor application.

More recently, government legislation mandated passive restraint systems for drivers of passenger vehicles. The manufacturers' choices were automatic seat belts or airbag systems. Customers quickly accepted the airbag system after numerous reports of users walking away from accidents that previously would have been classified as fatal crashes. As a result, an application that was opposed by vehicle manufacturers in the 1970s quickly became a selling feature in the 1990s. Buyers perceived an airbag system, relative to the cost of a vehicle, as a reason to buy one vehicle over another or, in some cases, justification for buying a new vehicle because their old vehicle just was not as safe.

The mechanical crash sensors initially used in these systems did not have a self-test feature and required calibration for each vehicle body style. Semiconductor-based accelerometers have allowed the number of crash sensors to be reduced from as many as five to only one or two, depending on the system's redundancy requirements. Furthermore, the system can be tuned to meet the requirements of a particular vehicle style after the sensor is in the module by programming the EEPROM in the microcontroller. Several techniques have been developed by sensor manufacturers that allow the accelerometer to be activated for self-testing each time the vehicle is started. The ability to perform the self-test and verify that this critical input to the system is capable of performing its designed function any time during the life of the vehicle is comforting to both the driver and the front seat passenger in vehicles equipped with those systems. After several deaths were attributed to airbag systems that did not discriminate the size or the position of the driver or the passenger, more sensors have been added, creating what have been called smart airbag systems. Once again, legislation and customer demands have driven semiconductor sensor development and acceptance.

The obvious question is where are the next opportunities for semiconductor-based sensors to meet a legislated requirement and/or customer demand. The answer is revealed, in part, by the technology that is already being developed: the follow-the-money approach. Providing additional safety or security and reducing energy consumption and emissions are behind many of the potential applications in today's sensor R&D laboratories. National laboratories in the United States and universities supported by corporate funding are the basis of many of the developments cited in this book. Government interest in commercializing R&D is providing manufacturers and universities the opportunity to work with national laboratories and take the next step with off-setting government funds.

Other areas that are growing rapidly are handheld portable products. Today's highly mobile society uses technology in a variety of portable applications. Memory storage could benefit from a MEMS-based 1-in disk drive. Finding the smart sensor applications in portable products could open new markets for those products.

Legislated requirements always provide guaranteed market and focus development efforts on a specific application. Legislation has been enacted for fugitive emissions, carbon monoxide detection in homes, trailers, and recreational vehicles, and reduced energy consumption in industrial, building, and home applications. Applications that address an aging population of baby boomers with discretionary funds to spend on entertainment and the avoidance of the effects of aging (e.g., hearing loss) are potential candidates for volume sensor usage. The question that must be answered now is can these semiconductor sensors provide cost-effective, value-added functionality to the system. That can occur in an application in which the sensor replaces a previous technology and allows more users to take advantage of an existing system based on the cost reduction. Another possibility is a new system that can be designed based on the capability provided by semiconductor sensors combined with other new technologies. Anything that combines smart sensors with the Internet is obviously attractive. The examples in Section 14.4 demonstrate research that could be the next enabling technology for sensors.

14.4 Not-So-Futuristic Systems

Many of the sensors that will be produced in the future will not surprise industry watchers because researchers will have demonstrated them at early stages of the technology development. Table 9.1 listed MEMS technology that could also produce manufacturable products at some point in the future. Examples in this section include a laboratory tool that can help researchers as well as technologies that may be used directly by a large number of consumers. The following is a list of some of the more intriguing sensors that are in development:

- "Nose-on-a-chip";
- Gas analysis system;
- Camera-on-a-chip;
- Atomic-force microscope;
- Monolithic hearing aid;
- Mass spectrometer;
- Vehicle dynamics sensor;
- Viscometer-on-a-chip;
- Chemistry lab-on-a-chip;
- DNA analyzer-on-a-chip;
- ID verification-on-a-chip.

14.4.1 Fabry-Perot Interferometer

Fabry-Perot interferometers (FPIs) are used in the laboratory to optically measure lengths or changes in length with a high degree of accuracy. A micromachined version consists of two silicon wafers with deposited highly reflective dielectric mirrors, as shown in Figure 14.1 [8]. This is already a second-generation version of an earlier development. An FPI incorporates a number of critical micromachined elements and semiconductor processes, including the controlled gap, movable silicon mesa corrugated support, silicon fusion bonding, capping wafer, control electrodes, and optical coating. Miniature FPIs have applications in telephony systems to link large numbers of customers to a central exchange. In addition, a small-gap FPI may have applications in sensors to measure small displacements and provide the feedback element in a servo loop. Two visible-ultraviolet-range interferometers in series are expected to give narrow-bandwidth selectivity for use as a spectrometer [9].

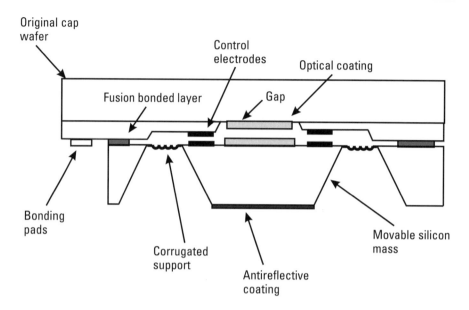

Figure 14.1 Fabry-Perot interferometer. (*After:* [8].)

14.4.2 HVAC Sensor Chip

Improved energy efficiency and energy management are major driving forces for new products. The energy consumption is directly achieved by using improved efficiency power devices and control techniques, such as PWM. The control techniques allow variable-frequency-drive motors to be implemented in building automation. Buildings in the United States account for 35–40% of the nation's energy consumption.

All the MEMS devices for an HVAC system eventually may be integrated into a single device, like the one shown in Figure 14.2 [10]. In addition to the sensors, signal conditioning to amplify the signal, calibrate offset and full-scale output, and compensate for temperature effects could be included on this device. Because it most likely will use CMOS technology for the semiconductor process, an onboard ADC also could be integrated. Other system capabilities could include fuzzy logic and neural network interpretation of the input signals. That is especially true if an array of chemical sensors is used to indicate a wide variety of chemical species and overcome many of the problems of available chemical-sensing products. Currently, sensors are not that highly integrated and are specified on an as-required basis. That is because of the additional cost they add to the initial system installation and their inability to function without requiring more maintenance than other parts of the system.

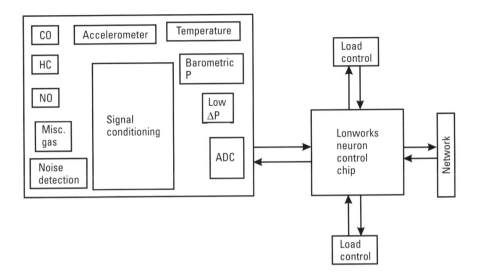

Figure 14.2 A highly integrated sensor concept for automated building control.

An example of how micromachining is extending the types of measurements and improving the performance compared to previous approaches is a MEMS-based chemical sensor. For those devices, a thin-film metal oxide technology is combined with an embedded microheater on a thin, micromachined silicon diaphragm. The small sample area of a micromachined sensor can more readily be raised to the higher operating temperature required for detecting the presence and actual value of specific chemicals. The polysilicon heaters in the chemical sensor operate at temperatures that can reach 450°C. The ability to change the temperature in a short period of time is used to minimize the effects of humidity and also to reduce power consumption [11]. Sensitivity from 10:1 to 100:1 for a specific gas and a response time of 2 minutes or less are common for these devices. These products are aimed at industrial and consumer applications, but ongoing development should provide units for automotive and other extremely cost-sensitive applications in the future.

14.4.3 Speech Recognition and Micromicrophones

Combinations of technologies will provide future products. One possibility is the combination of speech recognition algorithms, complex computing capability, and micromachined microphones. Neural networks are being investigated for intelligent sensors in speaker-independent speech-recognition systems [12]. Such research has resulted in an acoustic preprocessing chip that offloads the computational requirements from the host computer, allowing a slower, more cost-effective computer to perform speech recognition. That portion of the system, combined with a micromachined microphone like the one developed by the University of California, Berkeley, could provide portable speech recognition and possibly language translation capability [13]. Of course, both portions of these systems still require considerable development effort to be manufactured as separate elements. However, when they can be manufactured, the combination of devices and the resulting benefits will become obvious to many designers.

14.4.4 Electrostatic Mesocooler

The U.S. Defense Advanced Research Projects Agency (DARPA) is investigating what is being called mesoscale machines [14]. Mesoscale machines straddle the size range between MEMS and conventional machines. At some point, these devices may be implemented with MEMS technology. Figure 14.3 illustrates the concept of an electrostatic mesocooler in a refrigerated vest. The advantages of the existing approach are:

- One-third the weight of the conventional system;
- High coefficient of performance;
- Large Q due to high surface area (channels in parallel);
- Low power consumption (0.25–1W);
- Good flow rates;
- High compression ratio (4 atm).

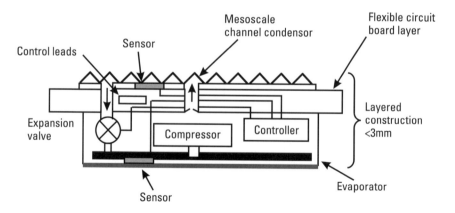

Figure 14.3 An electrostatic mesocooler.

Taking the next step using MEMS may be a logical extension of this technology.

14.4.5 Microangular Rate Sensors

Small cost-effective angular rate sensors are being developed for vehicle stability, adaptive cruise control, and vehicle navigation systems. One approach uses a plating mold for a metal structure for sensing angular rate [15]. As shown in Figure 14.4(a), the initial surface micromachined sensor used a ring structure. The center is held in place by four spring support arms attached to four mounting posts. The sensor measures rotation around the X- and Y-axes. The more recent ring-comb combination increases the signal and improves the temperature performance. The signal conditioning is performed by CMOS circuitry that can be integrated on the same chip as the angular rate sensor. This combination of micromachining and microelectronics allows the angular rate sensor to be embedded in a vehicle control system for use by consumers. This sensor will also be connected to a vehicle communication bus for use in other systems. As with other CMOS MEMS, the possibility of monolithically integrating the circuitry with the structure should be both possible and cost effective in the future.

14.4.6 MCU With Integrated Pressure Sensor

A monolithic microcontroller with a micromachined pressure sensor has been demonstrated [16]. Figure 14.5 shows the key elements of the design. The MCU-sensor combination is based on a widely available MCU architecture, with an integral bulk-micromachined sensor. Fabricating the chip is only one aspect of taking it from concept to reality. The packaging and testing challenges discussed in Chapter 10 also must be addressed [17]. Currently, a separate MCU and sensor packaged at a module level provides a more cost-effective solution. However, a specific application may require the space savings or performance advantages that the monolithic solution can provide. At that point, the level V sensor shown in Figure 1.7, the fully integrated sensing system, will become a reality.

An example of an application for this might be that of a side airbag sensor [18]. A pressure sensor inside the door panel of a car could be used to sense the change in pressure when the panel crumples under an impact. The ability to program the on-chip microcontroller will enable the auto manufacturer to embed the control algorithm inside the chip. To complete an entire system, only a mechanism for actuating the airbag needs to be added. That actuation

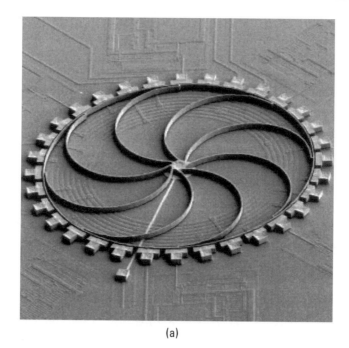

(a)

(b)

Figure 14.4 Micromachined (a) ring and (b) angular rate sensor. (Courtesy of Delphi Delco Electronic Systems.)

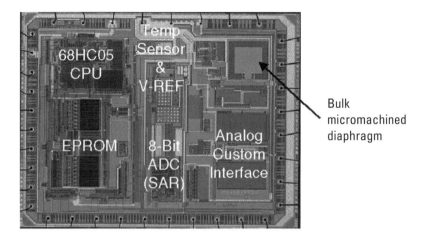

Figure 14.5 Photomicrograph of microcontroller with an integrated pressure sensor. (Courtesy of Motorola, Inc.)

capability could be another step in the continuous integration of silicon and electronic/electromechanical systems.

The proof of concept vehicle was built using a 100-kPa pressure sensor integrated onto a standard 8-bit 68HC05 MCU core along with the associated analog circuitry [18]. Figure 14.6 illustrates the block diagram, including an 8-bit MCU core (68HC05) with 2K bytes EPROM, 128 bytes RAM, boot ROM, and an SPI. Analog circuitry for signal conditioning, a voltage and current regulator, and a 10-bit ADC and 8-bit DAC were added to the basic core. A temperature sensor was also incorporated into the design for compensation purposes.

14.4.7 Wireless Sensing in the Networked Vehicle

RF technology can add another dimension to smart sensors. A vehicle provides some interesting possibilities. It is possible to envision small micromachined structures with an integrated RF output or input interfaced through an RF bus to a remote control unit performing their function without any visible means to verify their presence except the added functionality that they provide to the host vehicle. That technique is already used in electronic identification in the RFID tag (see Chapter 8). One possible vehicle application is a wireless version of an electronic "nose" [18]. The device could be mounted in an existing automotive component such as the dome light housing. The dome light housing provides a central position in the car for the nose to function, similar to a home

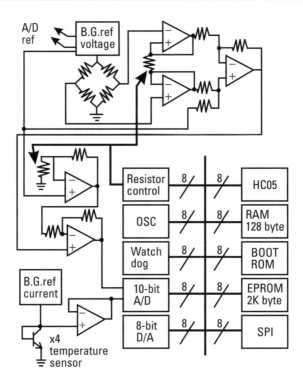

Figure 14.6 Block diagram of monolithic MCU-pressure combination.

smoke/chemical detector. Rather than sound an audible alarm, the sensor would send a wireless signal to a remote receiver/interpreting device in the vehicle. The signal would be received by an RF backbone with an appropriate protocol to recognize the particular sensor from other RF signals that could be transmitted in the vehicle.

The (wireless) sensory capability could be expanded to other defense mechanisms to detect and react to other harmful or undesirable occurrences in the vehicle. For example, the driver or a passenger could wear a sensing system that transmits the driver's physical state. People with potentially debilitating conditions, such as heart condition, epilepsy, or diabetes, would be prime targets for such sensing systems, adding to the total number of RF sensing nodes in the vehicle.

14.4.8 Personal ID Smart Sensor

Security is driving several approaches for automated personal identification [19]. The digital access of information and the ability to make transactions over the Internet are defining the need and value of such personal ID sensing, or biometrics. Potential users of biometrics are banks, governments, healthcare entities, and businesses. Sensors that recognize voice, face, fingerprints, hand, eye, and DNA are being investigated. Voice recognition is a behavioral identification; the other five are physiological. The combination of neural networks and sensing technology is being pursued to solve this smart sensing requirement, and a reliable, low-cost solution will be implemented for a number of applications.

14.5 Software, Sensing, and the System

CAD and simulation capability is essential in both the electrical and the mechanical portions of the sensor. It is also becoming a critical element in any system design. Design tools that implement mechatronic methodology are being standardized, which will produce a common level of understanding. Based on the increased capability that semiconductor hardware is providing, a more structured methodology will be the only way that designers will be able to cope with the semiconductor capability that will be available in the future. The number of transistors, the increased I/O possibilities, and the added difficulty inherent in high-density geometries will pose challenges to the most experienced designer. Add to that brief list only one more item, such as lower supply voltages, and it is easy to see that system designers must start to deal with system problems at higher and earlier stages in the design.

Software for sensor designers is already viewed as essential. Modeling for the structures and materials in the micromachining process has been the focus of universities and an integral part of the newest products provided by the sensor industry [20, 21]. Those activities, combined with mechatronics and the mixed-signal modeling that is gaining acceptance in the electronics design community, will provide the critical tools for smart sensing.

14.5.1 CAD for MEMS

Companies dedicated to providing tools and services for solving today's MEMS problems have been established and provide significant support to the development effort for MEMS. One company has developed an integrated design package that includes three-dimensional design, modeling, and simulation hardware [22]. This software supports electrostatics, electromechanics, microfluidics, thermodynamics, piezoresistance, and coupled thermomechanics. This MEMS-specific electronic design automation (EDA) tool integrates elecromechanical modeling with VHDL/Verilog and HDL-A support for mixed signal modeling. HDL-A is a high-level modeling language specific to MEMS. The software supports both bulk and surface micromachining and provides an etching verification and cross-section viewer.

As Chapter 10 pointed out, packaging is one of the major challenges that smart sensor manufacturers face. The three-dimensional simulation of the package by a MEMS tool can allow designers to quickly and easily build libraries of packages and device models that optimize the coupling of characteristics such as temperature-dependent effects [23]. For example, the temperature dependence of an accelerometer's package on the device can be demonstrated with available MEMS package software. Tools that have such capabilities should accelerate the transition from the laboratory to production and reduce the cost of the transition.

The future promise of MEMS sensors and actuators has prompted DARPA to invest roughly $50 million in the development of a composite CAD program for design tools [24]. The target devices include optical components such as tiny mirrors, vertical cavity lasers, and microfluidic components for chemical or biological testing. DARPA recognized that putting the electronics directly on a MEMS chip is "largely a research topic" with a few production examples such as accelerometers. They were interested in predicting the properties related to stress on the materials as well as the functionality of the electronic circuitry. Figure 14.7 shows DARPA's concept of the multiple steps involved in the design validation, synthesis, and simulation of a MEMS device [24]. The goal of the DARPA effort is to stay on the outer loop of the process shown in Figure 14.7 for most of the design process.

The need for software tools for the mechanical structures, electronic circuitry, and system characteristics that allow the interaction of critical system characteristics to be evaluated prior to fabrication is recognized and being advanced for smart sensors and MEMS-based systems in general. The combination of knowledge-based characterization of the system and standards provides the foundation to achieve world-class manufacturing capability for sensors. It is the manufacturing capability that will take sensors from the researchers' laboratories into the lives of consumers.

14.6 Alternative Views of Smart Sensing

The smart sensing definition in Chapter 1 is part of IEEE 1451. However, as the technology has evolved, other definitions have been offered that show a different view of smart sensing technology and offer insight to what researchers think is in the future for smart sensors. Two examples are presented here.

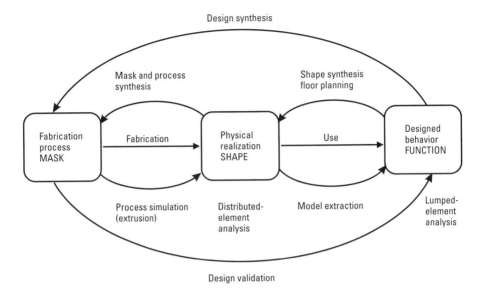

Design synthesis

Mask and process synthesis

Shape synthesis floor planning

Fabrication process MASK

Fabrication

Physical realization SHAPE

Use

Designed behavior FUNCTION

Process simulation (extrusion)

Distributed-element analysis

Model extraction

Lumped-element analysis

Design validation

Figure 14.7 Combined electromechanical and mixed signal IC modeling. (*After:* [22].)

One prior view of smart sensing proposed that a "smart sensor is defined as one that is capable of: (i) providing a digital output; (ii) communicating through bi-directional digital bus; (iii) being accessed through a specific address; and (iv) executing commands and logical functions" [25]. The smart sensor also has desirable functions such as compensation of secondary parameters (e.g., temperature), failure prevention and detection, self-testing, autocalibration, and various computationally intensive operations. The emphasis on computing attributes in this definition necessitates a microcontroller as a minimum requirement for a smart sensor.

Another proposed definition stated that "a smart sensor consists of a transducer combined with other signal conditioning and other circuitry to produce output information signals that have been logically operated on to increase the value of the information to the system" [26]. The aspects given in this definition of added signal conditioning and increased information value are noteworthy. The author lists sensor characteristics that take several sensors from the simplest to smartest version, as illustrated in Figure 14.8.

It is doubtful that either author of those two definitions would be satisfied with the definition proposed by the IEEE. However, the key to smart sensing will be in the ability to satisfy users, not those making definitions. Sensor

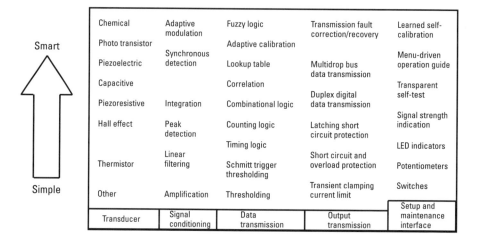

Figure 14.8 Smart sensor relative IQ. (*After:* [26].)

manufacturers will apply the term *smart sensor* to a product even if it does not meet an industry definition. But customers selecting the sensors that directly affect the performance and value of a product will determine the truly smart sensors. It is hoped that this book has in some manner provided guidance to manufacturers and users of smart sensors to understand their potential in control systems.

14.7 The Smart Loop

Smarter sensors require adding intelligence to the sensor. The easiest way to add intelligence is by using a microcontroller to provide computing capability and digital communication. Eventually, a monolithic combination of those technologies will be mass producible. In contrast, smarter systems require understanding more about the system and the environment. In real time, that knowledge must be accumulated from sensors that provide more information to the computing portion of the system, such as the smart airbag system mentioned earlier in this chapter. Figure 14.9 shows the continuum that exists as future sensors get smarter and sensors are added to controllers. Microsystems need smart sensors. Systems, in general, will benefit from smart sensors. If we start at any point on the continuum and add an increasing amount of smarts to sensors or more sensor inputs to the digital controller, the system moves the operating point from one perspective to the other. In either case, the system and users of the system benefit.

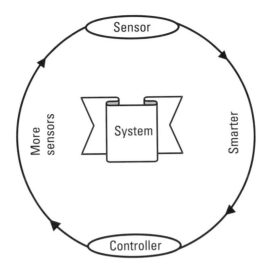

Figure 14.9 The smart loop.

14.8 Summary

Smart sensors will mean more than the IEEE definition presented in Chapter 1. However, technology is not smart if it is not cost effective. The exciting prospect of merged MEMS and microelectronics is a sensor that adds very little cost compared to the cost of all the other electronics required to make the sensor function as expected in the final system. The level V sensor, as proposed in Chapter 1, will use well-defined CMOS processing for all the microelectronics and a CMOS-compatible micromachining process for the sensor(s).

As noted in Chapter 1, semiconductor sensors originated as unwanted effects on transistors. As the industry improves the ability to fabricate micromachined sensors, especially surface micromachined sensors, the area of the sensor relative to circuitry that makes the sensor output useful and smarter becomes a smaller and smaller portion of the overall area. Additional processing for the sensor will increase the overall cost. Viewed from the perspective of a semiconductor manufacturer, the lion's share of the cost is in everything except the sensor. For all intents, the sensor is "free." The cost increases, sometimes significantly, as the specifications are defined for a specific sensor. Performance, especially accuracy, resolution, repeatability, and reliability, directly affect qualification efforts, test time, and yield—items that have a direct and significant impact on cost.

The cost is associated more with initial process, design, packaging, and, of course, testing of the sensing system. The cost of the CMOS portion will be reduced as the semiconductor industry's roadmap for improved capability is continually advanced. Achieving the desired specification will be where cost becomes a factor. A sensor used in a moderate (room temperature) environment with only 5% accuracy could have the same internal sensing chip as the rugged automotive application that achieves 1% accuracy over a −40°C to 125°C ambient temperature. The cost of testing, on-chip calibrating, and packaging will make the latter more expensive.

Unfortunately, the much-too-frequent prefix *smart* is used to describe anything that is an improvement over a previous version of a technology, product, or service. Examples of smart things include SmartTrading™ by OLDE, Smart Scrub™ from Dow, Smart Ones™ from Weight Watchers®, Smart-Rate℠ from Discover® Card, Smart Solutions℠ from the United States Postal Service, and even Smart, Very Smart™ products from Magnavox.[1] *Smart* is the late-twentieth-century term that was used instead of "new and improved." And there is every reason to expect that the term will carry over well into the twenty-first century.

Recent and projected advances in semiconductor technology and design tools will enable tomorrow's engineers to design, simulate, and verify something as complex as an entire vehicle system and probably to "virtually" prototype the entire vehicle. To effectively utilize those tools, the design methodology and even the designers' skill sets must be modified to include MEMS technology in the systems approach. Standards, such as IEEE 1451, will play a critical role in reducing costs for sensor manufacturers, savings that can be passed on to systems manufacturers and, ultimately, to consumers. Achieving increased performance and functionality at costs that were once thought to be impossible will make future smart sensors possible, limited only by the imagination.

References

[1]   Grace, R. H., "The Commercialization of MEMS/MST: Where Do We Go From Here?" *Sensors*, Feb. 1999, pp. 4–8.

[2]   Matsumoto, C., "Study calls MEMS Technology 'Disruptive' to Industry," http://www.techweb.com/se/directlink.cgi?EET19970804S0106

[3]   *The National Technology Roadmap for Semiconductors Technology Needs*, Semiconductor Industry Association, San Jose, CA, 1997.

---

1. All trademarks and service marks are the property of their respective owners.

[4] Studt, T., "Smart Sensors Widen View on Measuring Data," *R&D Magazine*, Mar. 1994, pp. 18–20.

[5] Haystead, J., "SIA Roadmap Highlights Contamination Control," *Cleanrooms*, Vol. 9, No. 2, Feb. 1995, pp. 1, 10.

[6] Stock, A. D., and D. R. Judd, "Interoperability Standard for Smart Sensors and Actuators Used in Semiconductor Manufacturing Equipment," *Proc. Sensors Expo*, Philadelphia, Oct. 26–28, 1993, pp. 203–222.

[7] Tepermeister, I., "Smart Sensors: The Next Evolutionary Step in Metrology," *Solid State Technology*, Mar. 1997, pp. 167–168.

[8] Jerman, J. H., D. J. Clift, and S. R. Mallinson, "A Miniature Fabry-Perot Interferometer With a Corrugated Silicon Diaphragm," *Technical Digest of IEEE Solid-State Sensor and Actuator Workshop*, Hilton Head Island, SC, June 4–7, 1990, pp. 140–144.

[9] Raley, N. F., et al., "A Fabry-Perot Microinterferometer for Visible Wavelengths," *Technical Digest of IEEE Solid-State Sensor and Actuator Workshop*, Hilton Head Island, SC, June 22–25, 1992, pp. 170–173.

[10] Frank, R., and D. Walters, "MEMS Applications in Energy Management," *Proc. Sensors Expo*, Boston, May 16–18, 1995, pp. 11–18.

[11] http/www.sandia.gov/media/acoustic.htm

[12] Donham, C., et al., "A Neural Network Based Intelligent Acoustic Sensor," *Proc. Sensors Expo*, Philadelphia, Oct. 26–28, 1993, pp. 121–127.

[13] Ried, R. P., et al., "Piezoelectric Microphone With On-Chip CMOS Circuits," *J. Micromechanical Systems*, Vol. 2, No. 3, Sept. 1993, pp. 111–120.

[14] http://www.darpa.mil/dso/darpa40/future/90Fmesocool.html

[15] Zarabadi, S., et al., "A Resonating Comb/Ring Angular Rate Sensor Vacuum Packaged Via Wafer Bonding," SAE 99011043, reprinted from *Sensors and Actuators 1999* (SAE SP-442), Internat'l Congress, Detroit, Mar. 1–4, 1999.

[16] Yoshii, Y., et al., "1 Chip Integrated Software Calibrated CMOS Pressure Sensor With MCU, A/D Convertor, D/A Convertor, Digital Communication Port, Signal Conditioning Circuit and Temperature Sensor," *Proc. Transducers '97*, Chicago, June 16–19, 1997.

[17] Frank, R., and D. Zehrbach, "Testing the System on a Chip," *Proc. Sensors Expo*, San Jose, CA, May 19–21, 1998, pp. 235–238.

[18] Frank, R. "Embedded Micromachines Can Eliminate Unnecessary Appendages," 98C041 *Proc. Convergence 96*, Oct. 21–23, 1996, pp. 349–354.

[19] Miller, B., "Vital Signs of Identity," *IEEE Spectrum*, Feb. 1994, pp. 22–30.

[20] Zhang, Y., S. B. Crary, and K. D. Wise, "Pressure Sensor Design and Simulation Using the CAEMEMS-D Module," *Technical Digest of IEEE Solid-State Sensor and Actuator Workshop*, Hilton Head Island, SC, June 4–7, 1990, pp. 32–35.

[21] Harris, R., M. F. Maseeh, and S. D. Senturia, "Automatic Generation of a 3-D Model of a Microfabricated Structure," *Technical Digest of IEEE Solid-State Sensor and Actuator Workshop*, Hilton Head Island, SC, June 4–7, 1990, pp. 36–41.

[22] Travis, B., "MEMS Sensors and Mechanical Gadgets Enter the Mainstream," *EDN*, May 21, 1998, pp. 77–87.

[23] Gilbert, J. R., and S. F. Bart, "Enabling the Design and Use of MEMS Sensors," *Sensors*, Apr. 1998, pp. 34–37.

[24] Goering, R., "Darpa Program Seeds Mixed-Technology CAD," http://www.eet.com/news/97/938news/seeds.html, Jan. 20, 1997.

[25] Najafi, K., "Smart Sensor," *J. Micromechanics and Microengineering*, June 1991, pp. 86–102.

[26] Juds, S. M., "Towards a Definition of Smart Sensors," *Sensors*, July 1991, pp. 2–3.

List of Acronyms and Abbreviations

3C-SiC	epitaxial silicon carbide
A/D	analog to digital
A/F	air-fuel ratio
A-bus	Automotive Bit-Serial Universal Interface System
ACK	acknowledge
ADC	A/D converter
ADPCM	adaptive differential pulse code modulation
ADS	automatic driver sensor
AFM	atomic force microscope
AHDL	analog hardware description language
Al	aluminum
ALU	arithmetic logic unit; the unit of a computing system containing circuits that perform arithmetic and logical operations
AMRA	Automatic Meter Reading Association
ANSI	American National Standards Institute
API	application programming interface

APK	amplitude-phase keying; a modulation technique for RF signal transmission
APTS	advanced public transportation system
ARCNet™	Attached Resource Computer Network
ARTS	advanced rural transportation system
ASHRAE	American Society of Heating, Refrigeration, and Air-Conditioning Engineers
ASIC	application-specific integrated circuit; an IC designed for a custom requirement, frequently a gate array or programmable logic device
ASK	amplitude-shift keying; a modulation technique for RF signal transmission
ASME	American Society of Mechanical Engineers
ASR	automatic speech recognition
ASTM	American Society for Testing and Materials
ATIS	advanced traveler information system
ATMS	advanced traffic management system
ATPG	automation test program/pattern generator
AVCS	advanced vehicle control system
BAW	bulk acoustic wave
BER	bit error rate
BGA	ball grid array
BiCMOS	combination of bipolar and CMOS
BIST	built-in self-test; design technique that allows a chip to be tested for a guaranteed level of functionality
BPSG	boro-phospho-silicate glass
C4	controlled collapse chip connection; solder bumps on IC for circuit connection
CAD	computer-aided design

CAIBE	chemically assisted ion beam etching
CAN	Controller Area Network
CARB	California Air Resource Board
CCD	charge-coupled device
CCITT	Consultive Committee of the International Telephone and Telegraph
CDMA	code division multiple access; spread spectrum method of allowing multiple users to share the radio frequency spectrum by assigning each active user an individual code
CDPD	cellular digital packet data; wide area data network that takes advantage of existing AMPS (U.S.) cellular network by transmitting data packets on unused voice channels. Data is transmitted using RS (63, 47) at an effective rate of 14 Kbit.
CEBus	consumer electronics bus
CERDIP	ceramic DIP
CISC	complex instruction set computer; standard computing approach as compared to RISC architecture
CMOS	complementary metal oxide semiconductor
CMR or CMRR	common mode rejection (ratio); the ratio of the common-mode input voltage to output voltage commonly expressed in dBV, that is, the extent to which a differential amplifier rejects an output when the same signal is applied to both inputs
CO	carbon monoxide
COB	chip on board; packaging technique for semiconductor die in multichip modules
codec	code/decode; translates audio to digital signals and digital back to audio signals, usually with an A/D and a D/A converter
COP	chip operating properly

CPU	central processing unit; the portion of the computer that includes the circuits that control the interpretation and execution of instructions
CRC	cyclic redundancy check; an error-detecting code in which the code is defined to be the remainder resulting from dividing the bits to be checked in the frame by a predetermined binary number
CSIC	customer-specified integrated circuit
CSMA	carrier sense, multiple access
CSMA/CD	carrier sense, multiple access with collision detection; access technique used in the Ethernet protocol
CSMA/CR	carrier sense, multiple access/collision resolution
CVO	commercial vehicle operation
DA	data acquisition
DAC	digital-to-analog converter
DARPA	Defense Advanced Research Projects Agency; U.S. agency investing in MEMS research as well as other advanced technologies
DCS	distributed control system
DFT	design for testability; methodology that takes test requirements into account early in the design process
DIN	Deutsches Industrial Norm, a set of technical/scientific and dimensional standards developed by a German organization
DIP	dual in-line plastic
DLC	data link controller
DMOS	diffused MOS device
DPSK	differential phase shift keying; modulation technique for transmission in which the frequency remains constant, but the phase changes from 90 degrees, 180 degrees, and 270 degrees to define the digital information

DQSK	differential quadrature shift keying
DRAM	dynamic random access memory
DSI	distributed systems interface
DSP	digital signal processing; a process by which a sampled and digitized data stream is modified to extract relevant information; also, a digital signal processor
DTS	digital termination service
DUT	device under test
DWB	direct wafer bonding
E-beam	electron-beam
ECG	electrocardiogram
EDA	electronic design automation
EDP	ethylene diamine pyrocatechol; an etchant for bulk micromachining
EEPROM	electrically erasable programmable read only memory; a semiconductor technique used for permanent storage that can be reprogrammed in the system
EFAB	electrochemical fabrication; a micromachining process that relies on the electroplating of metal and an in-situ mask patterning technique
EIA	Electronic Industries Association
EMC	electromagnetic compatibility
EMI	electromagnetic interference; see *RFI*
EOB	electrical oscillator-based
EPROM	erasable programmable read only memory; a semiconductor technique used for permanent storage but can be erased by ultraviolet light
ESD	electrostatic discharge; an electrical discharge usually of high voltage and low current
ESDA	electronic systems design automation

ETSI	European Telecommunications Standards Institute
EVM	evaluation module
FCC	Federal Communications Commission
FDMA	frequency division multiple access; method in which each active user in multiple user system is assigned an individual frequency channel
FEA	finite element analysis; mechanical simulation of stress
FED	field emission display
FeRAM/FRAM	ferroelectric random access memory
FET	field-effect transistor
FFT	fast Fourier transform; algorithm for reducing the number of calculations to extract frequency information from a signal
FIB	focused ion beam micromachining process
FIR	finite impulse response; filter whose output is determined by its coefficients and previous inputs. It is characterized by a linear phase response.
FIU	fuzzy inference unit
FM	frequency modulation; carrier wave modulation using frequency variation in proportion to the amplitude of the modulating signal
FMEA	failure-mode-and-effect analysis
FPAA	field-programmable analog array
FPGA	field-programmable gate array; an IC that can be programmed after it is assembled into a circuit
FPI	Fabry-Perot interferometer
FSK	frequency shift keying; a method of frequency modulation in which the modulating wave shifts the output frequency between predetermined values, and the output wave has no phase discontinuity

GaAs	gallium arsenide; a compound semiconductor material that has higher speed and higher operating temperature capability than silicon
GaN	gallium nitride
GF	gauge factor
GMSK	Gaussian minimum shift keying; a modulation technique for RF signal transmission
GPS	global positioning system
HART	highway addressable remote transducer
HC	hydrocarbons
HCMOS	high-density CMOS
HDL	hardware description language
HDL-A	hardware description language analog
HDLC	high-level data-link control
HF	hydrofluoric acid
HIV	human immunodeficiency virus
HLL	high-level language; a programming language that utilizes macro statements and instructions that closely resemble human language or mathematical notation to describe the problem to be solved or the procedure to be used
HMM	hidden Markov modeling
HTML	hypertext markup language
HVAC	heating, ventilation, and air conditioning
I/O	input/output for a semiconductor or circuit
I^{2}C	inter-IC
I^{2}L	integrated-injected logic
IC	integrated circuit; a semiconductor that has several hundred or more transistors designed into it
IDB	ITS databus

IEEE	Institute of Electrical and Electronics Engineers
IGEFET	interdigitated gate electrode field-effect transistor
IIR	infinite impulse response; filter whose output is determined by its coefficients, current output, and previous inputs. It is characterized by a nonlinear phase response.
IMEMS	intelligent MEMS
InP	indium phosphide
IP	Internet protocol
IPS	integrated pressure sensor
IR	infrared
$\overline{IRQ}$	interrupt request not
IS^2	Integrated Smart Sensor
ISA	Instrument Society of America
ISM	industrial, scientific, and medical frequency bands that were allocated by the FCC to spur rapid development of RF applications in a virtual open-market fashion. Licensing is automatic.
ISO	International Organization for Standardization
ISPF	Interoperable Systems Project Foundation
ITS	intelligent transportation system
IVHS	intelligent vehicle highway system; now ITS
JEDEC	Joint Electron Devices Engineering Council
JIS	Joint Industrial Standards. Also Japanese Industrial Standards Committee (JISC)
JPL	Jet Propulsion Laboratory
KGD	known good die; a specially tested die
KOH	potassium hydroxide
LAN	local area network
LCD	liquid crystal display

LDMOS	lateral diffused MOS
LED	light-emitting diodes
LGA	land grid array
Li/TiS$_2$	lithium titanium suphur
lidar	laser infrared radar
LIGA	process developed by the Fraunhofer Institute in Munich that combines X-ray lithography, electroforming, and molding to obtain high-aspect ratio micromachined structures. LIGA is derived from the German terms for lithography, electroforming, and molding.
LNA	low-noise amplifier
LOS	line of sight
LPCVD	low-pressure chemical vapor deposition
LSB	least significant bit
LWIM	low-power wireless integrated microsensor
MAC	media access control
Man	Manchester
MCAN	Motorola CAN
MCM	multichip module; the interconnection of two or more semiconductor chips in a semiconductor-type package
MCNC	Microelectronics Center of North Carolina
MCU	microcontrol unit or microcontroller unit; a semiconductor that has a CPU, memory, oscillator, and I/O capability on the same chip that is programmed to perform a limited number of functions
MDLC	Motorola's data link controller
MEMS	microelectromechanical system
MESFET	metal semiconductor field effect transistor; a high-frequency semiconductor device produced in GaAs semiconductor technology

MFM	modified frequency modulation
MIOS1	modular input/output subsystem
MIPs	millions of instructions per second; a measurement of microprocessor throughput
MJTC	multijunction thermal converters
MMIC	monolithic microwave integrated circuit; a high-frequency integrated circuit
modem	modulator-demodulator; unit that modulates and demodulates digital information from a terminal or computer port to an analog carrier signal for passage over an analog line
MOMS	microoptomechanical system
MOPS	millions of operations per second
MOS	metal oxide semiconductor
MOSAIC™	Motorola Self-Aligned IC
MOSFET	metal oxide semiconductor field effect transistor
MOSIS	Metal Oxide Semiconductor Implementation System
MPS	Michigan parallel standard
MPU	microprocessor unit; a central processing unit on a chip, usually without I/O or primary memory storage
MR	magnetoresistive
MRI	magnetic resonance imaging
MST	microsystem technology; the forming of structures using micromachining
MUMPS	multiuser MEMS process
MUX	multiplex; the combining of several messages for transmission over the same signal path
N_2H_4	hydrazine
NAN	niche area network

NaOH	sodium hydroxide
NASCAR	National Association of Stock Car Auto Racing
NCAP	network capable application processor; a device that supports a network interface, application functionality, and general access to the physical world via one or more transducers as defined in the IEEE 1451 standard
NiCd	nickel cadmium
NIST	National Institute of Standards and Technology in Gaithersburg, MD; formerly, the National Bureau of Standards
NMOS	n-channel metal oxide semiconductor
NODS	near-obstacle detection systems
NO_X	oxides of nitrogen
NRZ	nonreturn to zero; a data format in which the voltage or current value (typically voltage) determines the data bit value (one or zero)
NRZI	nonreturn-to-zero inverted
NSF	National Science Foundation
NTC	negative temperature coefficient; an NTC thermistor is one that the zero-power resistance decreases with an increase in temperature
NV	nonvolatile
OBDII	onboard diagnostics II; an automotive regulation defined by the California Air Resources Board
OLE	object linking and embedding
OMPAC	overmolded pad array carrier
OOK	on-off keying; the simplest form of ASK, in which the carrier is switched on and off by the pulse code modulation waveform
op amps	operational amplifiers
OPC Foundation	OLE for Process Control Foundation

OS	operating system
OSEK	German acronym for Open System for Automotive Electronics, an extension of the front end of the MCU hardware
OSI	Open Systems Interconnection
OSS	octal serial switch
OTP	one-time programmable
PAC	pad array carrier
PCB	printed circuit board
PCMCIA	Personal Computer Memory Card International Association
PGA	pin grid array
PIC	power IC
PID	proportional integral differential; control technique commonly used in servo systems
PLC	programmable logic controller
PLD	programmable logic device
PLL	phase locked-loop; a major component in a frequency synthesizer
PMMA	polymethyl methacrylate
Profibus	process fieldbus
PSG	phosphosilicate glass
PTC	positive temperature coefficient; a PTC thermistor is one that the zero-power resistance increases with an increase in temperature
PWM	pulse-width modulation; the digital control of voltage pulses used in amplifiers to obtain higher efficiency than linear control
PZT	lead zirconate titanate
QPSK	quadrature phase shift keying

QSPI	queued SPI
R&D	research and development
rad	radian
radar	radio detection and ranging; system based on transmitted and reflected RF energy for determining and locating objects, measuring distance and altitude, and navigating
RAM	random-access memory; memory used for temporary storage of data. RAM is volatile memory that is lost when the power is turned off.
RC	resistance-capacitance
RF	radio frequency; frequencies of the electromagnetic spectrum normally associated with radio wave propagation; sometimes defined as transmission of any frequency at which coherent electromagnetic energy radiation is possible, usually above 150 kHz
RFI	radio frequency interference; (usually) unintentionally radiated energy that may interfere with the operation of or even damage electronic equipment
RFIC	radio frequency IC
RF-ID	RF identification
RIE	reactive ion etching; a dry etching process that combines plasma etching and ion beam removal of the surface layer
RISC	reduced instruction set computer; a CPU architecture that optimizes processing speed by the use of a smaller number of basic machine instructions
rms	root mean square
ROM	read-only memory; memory used for permanent storage of data. ROM is nonvolatile memory.
RSD	remote sensing device
RTD	resistance temperature detector
RTR	remote transmission request

SAD	stochastic analog-to-digital
SAE	Society of Automotive Engineers
SAR	successive approximation register
SAW	surface acoustical wave
SBA	solder bump array
SCAP	silicon capacitive absolute pressure
SCI	serial communications interface
SCL	serial clock line
SCREAM	single-crystal reactive etching and metallization; a micromachining process that uses reactive ion etching
SDL	serial data line
SEM	scanning electron microscope
SEMI	Semiconductor Equipment Materials International
SERCOS	serial, real-time communication system
SFB	silicon fusion bonding
SHAL	Smart House Applications Language
SI (units)	International System of Units established in 1960 by the 11th General Conference on Weights and Measures and based on the metric system
Si	silicon
SIA	Semiconductor Industry Association
SiC	silicon carbide; a high-temperature semiconductor, sensor, and MEMS material
Si_3N_4	silicon nitride
SiO_2	silicon dioxide
SIOP	serial I/O port
SIP	single in-line plastic
SLIGA	LIGA process with the addition of a sacrificial layer

SMA	shape memory alloy
SMD	surface mount device; see *SMT*
SMT	surface mount technology; method of attaching components, both electrically and mechanically, to the surface of a conductive pattern
SNS	smart inertial navigation system
SOC	system on a chip; a highly integrated application-specific integrated circuit
SPI	serial peripheral interface
SPL	sound pressure level
sr	steradian
SRAM	static RAM
SSVS	super smart vehicle systems; term used in Japan for vehicles with several new electronic systems, typically used in ITS
STIM	smart transducer interface module; module defined in IEEE 1451 standard that includes the TEDS and the supporting electronics, including the transducer on the transducer side of the hardware interface to the NCAP
TAB	tape-automated bonding; semiconductor packaging technique that uses a tiny leadframe to connect circuitry on the surface of the chip to a substrate instead of wire bonds
TBC	transducer bus controller
TBIM	transducer bus interface module; module that consists of one to N transducers, an IEEE 1451.3-defined TEDS, and interface logic to control and transfer data across the minibus
Tco	temperature coefficient of offset
TCP	transmission control protocol
TCP/IP	transmission control protocol/Internet protocol
TCR	temperature coefficient of resistance

TDMA	time division multiple access; a technique that assigns each subscriber desiring service a different time slot on a given frequency. Signal compression is achieved by running at very high frequencies. Each user can then deliver the fixed packet message in a brief burst of time, thereby increasing the capacity of the system.
TEDS	transducer electronic data sheet; a machine-readable specification of the characteristics of the transducer interface defined in the IEEE 1451 standard
TEM	transverse electromagnetic mode
TII	transducer-independent interface; a 10-wire digital communication interface defined in IEEE 1451 standard that allows an NCAP or host to obtain sensor readings or actuator actions as well as request TEDS data
TiN/TiSi$_2$	titanium nickel/titanium silicon
TiW	titanium tungsten
TMAH	tetramethylammonium hydroxide
Tox	thermal oxide layer
TPU	time processor unit
TTP	time-triggered protocol
UART	universal asynchronous receiver transmitter
ULSI	ultra large scale integration; a chip with over 1,000,000 components
UPC	universal product code
USB	universal serial bus
VAN	vehicle area network
VAV	variable air volume
VCO	voltage-controlled oscillator; an oscillator whose output frequency varies with an applied dc control voltage
VFD	vacuum fluorescent display

VHDL	VHSIC hardware description language; a hardware description language for behavior-level circuit design developed by the U.S. Department of Defense
VHSIC	very high-speed integrated circuit
VLSI	very large scale integration; a chip with 100,000 to 1,000,000 components
VNAW	vehicle navigation and warning
vocoder	voice encoder
VPW	variable pulse width
VRML	virtual reality modeling language
W/E	write/erase
WIP	work in progress
WLAN	wireless LAN
WWW	World Wide Web
WZS	wireless zone sensors
ZnO	zinc oxide

Glossary

ablation   vaporization of material that occurs during laser trimming

accuracy   a comparison of the actual output signal of a device to the true value of the input. The various errors (such as linearity, hysteresis, repeatability, and temperature shift) attributing to the accuracy of a device are usually expressed as a percentage of full-scale output (FSO). Also see *error*.

actuator   the part of an open-loop or closed-loop control system that connects the electronic control unit with the process

adaptive (control)   refers to a system of advanced process control that is capable of automatically adjusting itself to meet a desired output despite shifting control objectives and process conditions or unmodeled uncertainties in process dynamics

algorithm   a set of well-defined rules or processes for solving a problem in a finite number of steps

aliasing   distortion due to sampling a continuous signal at too low a rate

aliasing noise   distortion component created when frequencies present in a sampled signal are greater than one-half the sample rate

analog output   an electrical output from a sensor that changes proportionately with any change in input

anisotropic   refers to etching that is dependent on crystallographic orientation

anneal   heat process used to remove stress especially in surface micromachining

antialiasing filter   normally, a low-pass filter that band limits an input signal before sampling to prevent aliasing noise

arbitration   process of gaining access to a bus

architecture   the hardware or software design, usually standardized, for a circuit or system; also, *system architecture*

associative memory   a neural network architecture used in pattern recognition applications, in which the network is used to associate data patterns with specific classes or categories it has already learned

attenuation   decrease in magnitude of communication signal

autozero   control circuitry that periodically reestablishes the zero to avoid drift errors

bandgap reference   forward-biased emitter junction characteristic of transistor used to provide an output voltage with zero temperature coefficient

bandpass filter   filter designed to transmit a band of frequencies while rejecting all others

baseband   frequency band occupied by information-bearing signals before combining with a carrier in the modulation process

baud   unit of signaling speed equal to the number of discrete signal conditions or events per second. Refers to the physical symbols per second used in a transmission channel.

biometrics   personal ID sensing

bit rate   speed at which data bits are transmitted over a communication path, usually expressed in bits per second (bps). A 9,600-bps terminal is a 2,400-baud system with 4 bit/baud.

bridge   device that forwards packets of information between channels in a multiple-media network

broadband   a modulated or square wave signal on a multiple modulated carrier

bulk micromachining   a process for making microstructures in which a masked (silicon) wafer is etched in orientation-dependent etching solutions

bus   connection between internal or external circuit components

calibration   process of modifying sensor output to improve output accuracy

carrier band   single-channel broadband; digital or modulated signal on a single modulated carrier

chip   a die (unpackaged semiconductor device) cut from a silicon wafer and incorporating semiconductor circuit elements such as a sensor, actuator, resistor, diode, transistor, and/or capacitor

closed loop   control system that utilizes a sensing device for measuring a process variable and making control decisions based on that feedback

collision   state of a data bus in which two or more transmitters are turned on simultaneously to conflicting states

combinational technologies   integrated mixed-signal (analog and digital) technology

companding   the process of reducing the dynamic range of voice or music signal (compressing) typically at the transmitter and then restoring (expanding) at the receiver

compensation   added circuitry or materials designed to counteract a known source of error

contention   ability to gain access to the bus on a predetermined priority

convolution   mathematical process that describes the operation of filters

coriolis force   force that creates a deflection of a body in motion with respect to the earth caused by the rotation of the earth; appears as a deflection to the right in the northern hemisphere and a deflection to the left in the southern hemisphere

data compression   technique that provides for the transmission of fewer data bits than originally required without information loss. The receiving location expands the received data bits into the original bit sequence.

data logging   method of recording a process variable over a period of time

decimation   process by which a sampled and digitized data stream is modified to extract relevant information

defectivity   in semiconductor wafer manufacturing, the defects per unit area

delta-P ΔP, change in pressure or pressure differential

deterministic   refers to the property of a signal that allows its future behavior to be precisely predicted

die   See *chip*

diffusion   a thermochemical process whereby controlled impurities are introduced into the silicon to define the piezoresistor. Compared to ion implantation, it has two major disadvantages: (1) The maximum impurity concentration occurs at the surface of the silicon, rendering it subject to surface contamination and making it nearly impossible to produce buried piezoresistors; and (2) control over impurity concentrations and levels is about one thousand times poorer than that obtained with ion implantation.

digital output   transducer output that represents the magnitude of the measurand in the form of a series of discrete quantities coded in a system of notation

direct sequence   fixed frequency spread spectrum technique

dissipation constant   the ratio (in milliwatts per degree Celsius) at a specified ambient temperature, of a change in power dissipation in a thermistor to the resultant body temperature change

Doppler effect   the change in the observed frequency of a wave caused by a time rate of change in the distance between the source and the point of observation

Doppler radar   radar that exploits the Doppler effect to measure the radial component of the relative velocity between the radar system and the target

down converter   a device that provides gain and frequency translation to a lower frequency

drift   an undesired change in output over a period of time, with constant input applied

dynamic range   the ratio of the largest to the smallest values of a range, often expressed in decibels

embedded control   See *embedded system*

embedded sensing   the use of a sensor in a product for a secondary instead of a primary function, for example, protection in a power device

embedded system   a system with one or more computational devices (which may be microprocessors or microcontrollers) that are not directly accessible to the user of the system

end point straight line fit   the maximum deviation of any data point on a sensor output curve from a straight line drawn between the end data points on the output curve

engine   the computational portion of an IC

epitaxial or epi   a single-crystal semiconductor layer grown on a single-crystal substrate and having the same crystallographic characteristics as the substrate material

error   the algebraic difference between the indicated value and the true value of the input; usually expressed in percentage of full-scale span and sometimes in percentage of the sensor output reading

error band   the band of maximum deviations of the output values from a specified reference line or curve due to those causes attributable to the sensor. Usually expressed as a positive percentage of full-scale output. The error band should be specified as applicable over at least two calibration cycles, so as to include repeatability, and verified accordingly.

etch stop   a layer of P-type material that etches at a much slower rate than N-type material

Ethernet   local area network software standard that determines system operation

excitation voltage (current)   the external electrical voltage and/or current applied to a sensor for its proper operation (often referred to as the supply circuit or voltage). See *supply voltage (current)*

feedforward   a method to precompensate a control loop for known errors

ferroelectric effect   the tendency of dipoles within a crystal to align in the presence of an electric field and to remain polarized after the field is removed

fieldbus   a two-way communications link among intelligent field devices and control systems devices

filter   a circuit that reduces noise and other unwanted elements of a signal

firmware   computer program or instructions stored in ROM instead of in the software

flash   semiconductor memory that can be used for permanent storage and that is easily electrically reprogrammed in the system; faster than EEPROM

floating point   representation of numbers in scientific notation with the exponent and the mantissa given separately to accommodate a wide dynamic range

flops   floating-point operations per second; a measurement of microprocessor performance

frame   a group of bits sent serially over a communications channel. The logical transmission unit sent between data link layer entities contains its own control information for addressing and error checking.

frequency hopping   in spread spectrum approach, the case of both base and subscriber (or handset and base) hopping from frequency to frequency simultaneously to minimize data lost through interference

full-scale output   the output at full-scale measurand input at a specified supply voltage; the sum of the offset signal plus the full-scale span

full-scale span   the output variation between zero and when the maximum recommended operating value is applied

fuzzy logic   a branch of logic that uses degree of membership in sets rather than a strict true/false membership

gateway   a node used to connect networks that use different protocols; a protocol converter

genetic algorithms   guided stochastic search techniques that utilize the basic principles of natural selection to optimize a given function

Harvard architecture   architecture in which on-chip program and data are in two separate spaces and are carried in parallel by two separate buses

hysteresis   a transducer's ability to reproduce the same output for the same input, regardless of whether the input is increasing or decreasing; also see *temperature hysteresis*

implant   See *ion implantation*

infrared   area in the electromagnetic spectrum ranging from 1 to 1,000 μm

input impedance (resistance)   the impedance (resistance) measured between the positive and negative (ground) input terminals at a specified frequency with the output terminals open

instrument   a system that allows measured quantity to be observed as data. An instrument consists of a sensing element and conversion, manipulation, and data transmission and presentation elements. All elements are not necessarily present in any particular instrument.

integrated refers to the combined design and fabrication of interconnected components. An integrated circuit contains a multiplicity of transistors as well as diodes, resistors, sensor(s), and so on.

integration the combination of previously discrete or separate circuit designs

interchangeability the ability of multiple systems or devices to exchange information and mutually use the information that has been exchanged between systems and devices

interoperability the capability of a computing or control system to replace component parts or devices with component parts or devices of different manufacturers and different product families, while maintaining the full range (or partial range) of optimal system functionality

ion beam milling a dry etching process that uses an ion beam to remove material through a sputtering action

ion implantation a process whereby impurity ions are accelerated to a specific energy level and impinged on the silicon wafer. The energy level determines the depth to which the impurity ions penetrate the silicon. Impingement time determines the impurity concentration.

isotropic etching that is independent of crystallographic orientation

kernel software that is embedded in the controller

laser trimming (automated) method for adjusting the value of thin- or thick-film resistors using a computer-controlled laser system

latency guaranteed access (with maximum priority) within a defined time

linearity error the maximum deviation of the output from a straight-line relationship input over the operating measurand range. The type of straight-line relationship (end point, least squares approximation, etc.) should be specified.

load impedance the impedance presented to the output terminals of a sensor by the associated circuitry

Loran C third-generation long-range navigation system that establishes position by measuring the time differences of signals arriving from fixed location transmitters

Manchester   a digital signaling technique in which there is a transition in the middle of each bit time. A 1 is encoded with a high level during the first half of the bit time, and a 0 is encoded with a low level during the second half of the bit time.

maximum operating temperature   the maximum body temperature at which the sensor will operate for an extended period of time with acceptable stability of its characteristics. This temperature is the result of the internal or external heating or both and should not exceed the maximum value specified.

maximum power rating   the maximum power that a sensor will dissipate for an extended period of time with acceptable stability of its characteristics

mechatronics   the synergistic combination of precision mechanical engineering, electronic control, and the systems approach for designing products and manufacturing processes

media   means of data transmission in a network, for example, two-wire, twisted-pair, coaxial cable, power lines, single-wire (with common ground), infrared, RF, fiber optic; also, the material used to transfer a pressure signal

mesoscale   a scale of machines that is between MEMS and conventional machines

microcode   machine instructions built permanently into controller circuitry

microcontroller   a single integrated circuit that contains a CPU, memory, a clock oscillator, and I/O. Also see *MCU* in the List of Acronyms and Abbreviations.

microengine   the computational portion of an IC

micromachining   the chemical etching of mechanical structures in silicon or other semiconductor material, usually to produce a sensor or actuator

micron 10^{-6} meter

mixed signal   the combination of analog and digital circuitry

mixer   device that utilizes nonlinear characteristics to provide frequency conversions from one frequency to another

monotonicity   a measurement of linearity. A monotonic curve is one in which the dependent variable either always increases or decreases as the independent variable.

multiplexer   device that allows two or more signals to be transmitted simultaneously on a single carrier or channel

multiprocessing   the simultaneous execution of two or more instructions by a computer

nanometer 10^{-9} meter

neural network   a collection of independent processing nodes that communicate with one another in a manner analogous to the human brain

Neuron™   a VLSI component that performs the network and application-specific processing within a node

node   any subassembly of a multiplex system that communicates on the signal bus

null offset   the electrical output present when the sensor is at null

null output   See *zero offset*

null temperature shift   the change in null output value due to a change in temperature

Nyquist rate   lowest sampling rate necessary to completely reconstruct a signal without distortion due to aliasing; equal to twice the highest frequency component in the signal

object   a collection of data and operations

object model   a definition of data structures and operations organized in a formal specification

observer   algorithm that corrects for variations from the physical model

offset   See *zero offset*

open loop   system with no sensory feedback

operating impedance   the impedance measured between the positive and negative (ground) output terminals at a specific frequency with the input open

operating temperature range   the range of temperature between minimum and maximum temperature at which the output will meet the specified operating characteristics

OSI model   the open system interconnect model of the ISO defines a seven-layer model for a data communications network

parallel computing   the processing of two or more (often many more) programs at one time by interconnected processing systems

partitioning   system design methodology that determines which portion of the circuit is integrated using a particular silicon process instead of completely integrating design using one process

peripherals   external circuit components necessary to achieve desired functionality, usually for an MCU or an MPU

photolithography   process in which a pattern in a mask is transferred to a wafer, resulting in areas to be doped or selectively removed

piezoelectric effect   the ability of certain materials to become electrically polarized in response to applied strain or to be strained in response to applied voltage

piezoresistor   a resistor that changes resistance in response to applied strain

pipeline   bus structure within an MPU that allows concurrent operations to occur

plasma etching   an etching process that uses an etching gas instead of a liquid to chemically etch a structure

polysilicon   silicon composed of randomly arranged crystal unit cells

precision   the smallest discernable change in the output signal of a device

protocol   the rules governing the exchange of data between networked elements

quadrature   a technique used in encoders and resolvers that separates signal channels by 90 degrees (electrical) in feedback devices

quantization error   the uncertainty $\pm 1/2$ least significant bit (LSB) that adds to conversion error in an analog to digital converter

range   See *operating temperature range*

ratiometric (ratiometricity error)   the change in the sensor output resulting from any change to the supply voltage (usually expressed as percentage of full-scale output). (At a given voltage, sensor output is a proportional value of that supply voltage.)

real-time system   system that accepts external inputs, processes them, and produces outputs in a fixed period of time

repeatability   the maximum change in output under fixed operating conditions over a specified period of time

resistance ratio characteristic   the ratio of the zero-power resistance of a thermistor measured at 25°C to that resistance measured at 125°C

resistance temperature characteristic   the relationship between the zero-power resistance of a thermistor and its body temperature

resolution   the maximum change in pressure required to give a specified change in output

response time   time required for a sensor output to change from its previous state to a final state within an error tolerance band of the new correct value

router   device that connects the channels in multiple-media control systems and passes information packets back and forth

sacrificial layer   a thin film deposited in the surface micromachining process that is later etched away to release a microstructure

scalable   ability of MCU or MPU architecture to be modified to meet the needs of several applications providing competitive price-performance points

Schmitt trigger   a circuit that turns the pulsed output from a sensor such as an optodetector or phototransistor into a pure digital signal. The lower and upper thresholds in the trigger remove the linear transition region between the on and off states.

self-generating   providing an output signal without applied excitation such as a thermoelectric transducer

self-heating   internal heating resulting from electrical energy dissipated within the unit

Sematech   consortium of United States semiconductor and equipment manufacturers

semiconductor sensor   sensor manufactured using silicon, GaAs, SiC, or other semiconductor materials

sensactor   a linguistic amalgam proposed by Ford engineers of a combined sensor, microprocessor, and actuator possibly integrated on a single silicon chip

sensitivity   the change in output per unit change in input for a specified supply or current

sensitivity shift   a change in sensitivity resulting from an environmental change, such as temperature

sensor   a device that provides a useful output to a specified measurand

servo mechanism   an closed-loop motion control system that uses feedback to control a desired output

set point   the desired value for a process (controlled) variable

silicon compiler   a tool that translates algorithms into a design layout for silicon

silicon fusion bonding   a process for bonding two silicon wafers at the atomic level without applying glue or an electric field; also called direct wafer bonding

simulation   design approach using computer models to predict circuit or system performance

skew rate   difference in delay between parallel paths

slew rate   maximum rate of change of voltage with time

smart power (ICs)   hybrid or monolithic devices that are capable of being conduction cooled, perform signal conditioning, and include a power control function such as fault management and/or diagnostics. The scope of [this definition] shall apply to devices with a power dissipation of 2W or more, capable of operating at a case temperature of 100°C and with a continuous current of 1A or more (JEDEC definition).

smart sensor   sensor that provides functions beyond those necessary for generating a correct representation of a sensed or controlled quantity

span   the output voltage variation given between zero and any given measurand input

SPECmarks   a normalized measurement of performance for RISC systems based on a standard set of operations

spectral density   the power at a specific frequency

spectral filter   a filter that restricts the electromagnetic spectrum to a specific bandwidth

spin-on   semiconductor process to deposit films and coatings

spread spectrum technique used to reduce and avoid interference by taking advantage of statistical means to send a signal between two points. A figure of merit for spread spectrum systems is "spreading gain" measured in decibels. The two types of commercial spread spectrum techniques are frequency hopping and direct sequence.

sputter semiconductor process to deposit a thin film of material, typically a metal, on the surface

squeeze-film damping effect of ambient gases and spacing on the movement of micromachined structures

stability the ability of a sensor to retain specified characteristics after being subjected to designated environmental or electrical test conditions

standard reference temperature the thermistor body temperature at which nominal zero-power resistance is specified (e.g., 25°C)

state machine logic circuitry that, when clocked, sequences through logical operations and can be a preprogrammed set of instructions or logic states

stiction static friction; the adhering of surface micromachined layers due to capillary forces generated during wet etching

storage temperature range the range of temperature between minimum and maximum that can be applied without causing the sensor (unit) to fail to meet the specified operating characteristics

submicron measurement of the geometries or critical spacing used for complex, highly integrated circuits

superscalar the ability of an MPU to dispatch multiple instructions per clock from a conventional linear instruction stream

supply voltage (current) the voltage (current) applied to the positive and negative (ground) terminals

surface micromachining a process for depositing and etching multiple layers of sacrificial and structural thin films to build complex microstructures

telemetry remote measurement that permits data to be interpreted at a distance from the detector

temperature coefficient of full-scale span the percentage change in the full-scale span per unit change in temperature relative to the full-scale span at a specified temperature

temperature coefficient of offset (Tco)   the percentage change in the offset per unit change in temperature relative to the offset at a specified temperature

temperature coefficient of resistance   the percentage change in the dc input impedance per unit change in temperature relative to the dc input impedance at a specified temperature

temperature error   the maximum change in output at any input value in the operating range when the temperature is changed over a specified temperature range

temperature hysteresis   the difference in output at any temperature in the operating temperature range when the temperature is approached from the minimum operating temperature and when approached from the maximum operating temperature with zero input applied

temperature-wattage characteristic   the relationship, at a specified ambient temperature, between the thermistor temperature and the applied steady-state wattage

thermal compensation   See *compensation*

thermal offset shift   See *temperature coefficient of offset*

thermal span shift   See *temperature coefficient of full-scale span*

thermal time constant   the time required for a thermistor to change to 63.2% of the total difference between its initial and final body temperatures when subjected to a step function change in temperature under zero-power conditions

thermal zero shift   See *temperature coefficient of offset*

thermistor   a thermally sensitive resistor whose primary function is to exhibit a change in electrical resistance with a change in body temperature

thin film   a technique using vacuum deposition of conductors and dielectric materials onto a substrate (frequently silicon) to form an electrical circuit

time constant   time required for a voltage or current in a circuit to rise to 63% of its steady final value or to fall to 37% of its initial value

topology   method of connecting data lines to system nodes

torr   pressure unit equivalent to 1 mm of mercury (mmHg)

transceiver   the combination of radio transmitter and receiver, usually using some common circuitry, in a portable or mobile application

transducer   a device converting energy from one domain into another and calibrated to minimize errors in the conversion process

transfer function   a mathematical, graphical, or tabular statement of the influence that a system or element has on the output compared at the input and output terminals

transponder   a transmitter-receiver that transmits signals automatically when a triggering signal is received

tribology   the study of friction, wear, and lubrication in surfaces sliding against each other, as in bearings and gears

twisted-pair Ethernet (10baseT)   Ethernet with a twisted-pair physical medium capable of carrying data at 10 Mbps for a maximum distance of 185m

upconverter   an upmixer with amplification and possibly other functions, such as power control and/or transmit envelope shaping

Verilog   a hardware description language for behavior-level circuit design placed in the public domain by Cadence Design Systems, Inc.

via   vertical length of metal deposited through a small hole in oxide used to electrically connect two layers of metal

Von Neumann (architecture)   refers to program and data being carried sequentially on the same bus

wafer   a thin, usually round slice of semiconductor material from which chips are made

X-ray (photo)lithography   technique used to achieve small physical spacing in integrated circuits

Young's modulus   modulus of elasticity

zero offset   the output at zero input for a specified supply voltage or current

Selected Bibliography

Books and Journals

Ko, W. H., and C. D. Fung, "VLSI and Intelligent Transducers," *Sensors and Actuators*, 2 (1982) pp. 239–250.

Middelhook, S., and A. C Hoogerwerf, "Smart Sensors: When and Where?" *Sensors and Actuators*, 8 (1985), pp. 39–48.

Muller, R. S., et al. (eds.), *Microsensors*, Piscataway, NJ: IEEE Press,1991.

Trimmer, W., *Micromechanics and MEMS*, Classic and Seminal Papers to 1990, Piscataway, NJ: IEEE Press, 1997.

Microelectromechanical Systems Advanced Materials and Fabrication Materials, Washington, D.C.: National Academy Press, 1997.

Sze, S. M., *Semiconductor Sensors*, New York: Wiley-Interscience, 1994.

Pallas-Areny, R., and J. G. Webster, *Sensors and Signal Conditioning*, New York: Wiley-Interscience, 1993.

Seippel, R. G., *Transducer Interfacing: Signal Conditioning for Process Control*, Reston, VA: Reston Publishing Co., 1988.

MindShare, Inc., and D. Anderson, *Universal Serial Bus System Architecture*, 2nd Ed., Reading, MA: Addison-Wesley, 1997.

Baxter, L. K., *Capacitive Sensors: Design and Applications*, Piscataway, NJ: IEEE Press, 1996.

Madou, M. J., and S. R. Morrison, *Chemical Sensing With Solid State Devices*, Boston: Academic Press, 1997.

Carr, J. J., *Sensors*, Vol. 1 of *Electronic Circuit Guidebook*, Indianapolis, IN: Prompt Publications, 1997.

MindShare, Inc., and D. Anderson, *FireWire System Architecture*, Reading, MA: Addison-Wesley, 1998.

Madou, M., *Fundamentals of Microfabrication*, Boca Raton, FL: CRC Press, 1997.

Taylor, R. F., and J. S. Schultz (eds.), *Handbook of Chemical and Biological Sensors*, Bristol, U.K.: Institute of Physics Publishing, 1996.

Yager, R. R., et al. (eds.), *Fuzzy Sets and Applications: Selected Papers by L. A. Zadeh*, New York: Wiley, 1987.

Kosko, B., *Neural Networks and Fuzzy Systems*, Englewood Cliffs, NJ: Prentice Hall, 1990.

Muller, N. J., and R. P. Davidson, *LANs to WANs: Network Management in the 1990s*, Norwood, MA: Artech House, 1990.

Greenberg, E., *Network Application Frameworks—Design and Architecture*, Reading, MA: Addison-Wesley.

Web Sites

University Web Sites

Berkeley Sensor and Actuator Center, http://www-bsac.eecs.berkeley.edu/

The Microelectromechanical Systems (MEMS) Resource at Case Western Reserve University, http://mems.eeap.cwru.edu/

Delft University of Technology, Micromachined adaptive mirrors, Gleb Vdovin Laboratory of Electronic Instrumentation, http://guernsey.et.tudelft.nl/tyson4/tyson4.html

Georgia Tech MEMS, http://mems.mirc.gatech.edu/

MCNC MEMS Technology Applications Center, http://mems.mcnc.org/mems.html

MIT Microsystems Technology Laboratories, http://goesser.mit.edu/

Northeastern University, Microfabrication Laboratory at Northeastern University, http://www.ece.neu.edu/edsnu/zavracky/mfl/mfl.html

Twente University Micromechanical Transducers Group Mesa Institute, http://www.el.utwente.nl/tdm/mmd/index.html

UCLA Wireless Integrated Network Sensors (WINS), http://www.janet.ucla.edu/WINS/

University of Michigan MISTIC (Michigan Synthesis Tools for ICs), http://www.eecs.umich.edu/mistic/

University of Wisconsin MEMS, http://mems.engr.wisc.edu/

Industry Web Sites

Echelon Corporation, http://www.echelon.com/

MEMS Optical Inc., http://www.memsoptical.com/mems/mems.htm

FlexNet, www.actionio.com

Government Web Sites

Sandia MEMS, http://www.mdl.sandia.gov/Micromachine/

DARPA, http://www.arpa.mil/mto/mems/

NIST 1451 Demonstration, http://motion.aptd.nist.gov/P1451/ISADemo.htm

Publishers and Events Web Sites

Sensors Magazine, http://www.sensorsmag.com/

Journal of Microelectromechanical Systems, http://home.earthlink.net/ ~trim-merw/mems/jmems.html

VDI/VDE-IT Microsystem Technology, http://www.vdivde-it.de/mst/

IEEE Sensors Council, International Conference on Solid-State Sensors and Actuators (TRANSDUCERS 'XX), http://www.ieee.org/web/search/

Control, http://www.controlmagazine.com/

Control Engineering, http://www.manufacturing.net/magazine/ce/

I&CS, http://www.icsmagazine.com/

Sensor-circuit integration and system partitioning (Kensall D. Wise), http://144.126.176.216/mems/d_5.htm

Elsevier Science Sensors and Actuators A/B, http://www.elsevier.com/

Association Web Sites

AS-Interface, http://www.as-interface.com/

Bitbus European Users Group, http://www.bitbus.org/wordofp.html

CEBus Industry Council, Inc., www.cebus.org

Controller Area Network (CANbus), an introduction to the serial communications business, http://www.ic.ornl.gov/p1451/whats_new_page.html

ControlNet International, http://www.controlnet.org/

DeviceNET, http://www.ab.com/networks/devicenet.html

Open DeviceNet Vendor Association, Inc., http://www.odva.org/

EIB (Home and Building Automation and Electrical Installation), www.eiba.be

Fieldbus Online, http://www.fieldbus.org/

HART Communication Foundation, http://www.ic.ornl.gov/p1451/whats_new_page.html

IEEE 1451, http://www.ic.ornl.gov/p1451/p1451.html

Interbus-S, www.interbusclub.com

Instrument Society of America, http://www.isa.org/index/

LonMark International, http://www.lonmark.org/

PLCopen, http://www.plcopen.org:80/

Profibus, www.profibus.com

SDS, http://www.honeywell.com/sensing/prodinfo/sds/

Semiconductor Industry Association, http://www.semichips.org

SERCOS, www.sercos.com

SwiftNet (SHIP STAR Associates), www.shipstar.com/swiftnet.html

OSI on WWW, http://www.ic.ornl.gov/p1451/whats_new_page.html

WorldFIP (World Factory Information Protocol), www.worldfip.org

Miscellaneous Web Sites

MiCroSystem Group Research Topics, http://tima-cmp.imag.fr/tima/mcs/topics.html

Microelectromechanical Systems Advanced Materials and Fabrication Methods, http://www.nap.edu/readingroom/books/micro/

MEMScaP Simulation tools for MEMS, http://memscap.e-sip.com/

MEMS Clearinghouse, http://mems.isi.edu/

MEMS, fabrication, consulting, micromechanics, MST, microelectromechanical systems, MEMS journals, MEMS books, MEMS clubs, MEMS Web sites, employment and jobs in the United States, text online, MEMS introduction, and MEMS conferences, http://home.earthlink.net/~trimmerw/mems/index.html

European Microsystems Technology Online, http://www.nexus-emsto.com/

Tutorial information about MEMS, http://mems.isi.edu/archives/otherWWWsites_tutorial.html

About the Author

Randy Frank is a technical marketing manager for ON Semiconductor in Phoenix, Arizona. He is also an application specialist in sensor technology. He has a BSEE, an MSEE, and an MBA from Wayne State University in Detroit and over 25 years of experience in automotive and control systems engineering, including the development of electronic engine controls and smart-power semiconductor products. Mr. Frank has been the chairman and is currently a member of the Sensor Standards Committee of the Society of Automotive Engineers (SAE). He is a member of the IEEE Standards Association and also a former member of the IEEE Sensor Terminology Taskforce and the Association for the Advancement of Medical Instrumentation (AAMI) Blood Pressure Transducer Committee. He has taught advanced instrumentation and control at the University of Michigan, including the reliability aspects of sensors. Mr. Frank holds three patents and has published over 200 technical papers and articles and several book chapters on semiconductor products and applications.

Index

in chemical measurements, 262
  gas sensors, 263
  infrared focal plane array, 263–64
  multichannel probe, 262
  photodiode arrays, 263
Multiple-type sensors, 264–65
Multiplexing (MUX), 120
Multiprocessor computing, 164
Multiuser MEMS process (MUMPS), 219

Nanoguitar, 222–23
N-channel metal oxide semiconductor
    (NMOS), 30
Near-obstacle detection systems
    (NODSs), 189–90
Network-capable application processor
    (NCAP), 274
  IEEE 1451.1, 276–78
  IEEE P1451.4, 293, 294
  PC, 281
  triggers, 284
Networked vehicle, wireless sensing, 323–24
Networks
  airplane, 306
  automotive safety, 306–8
  extending sensing systems to, 293–95
  industrial, 133
  neural, 157–60
Neural networks, 157–60
  application example, 160
  defined, 157–58
  fuzzy logic combined with, 160
  illustrated, 159
  neurons, 158
  uses, 158
  See also Control
Neuron IC, 139–41
  block diagram, 141
  defined, 139
  direct-mode transceiver, 140
Niche area network (NAN), 307
Noise, 65–66
  flicker, 66
  shot, 65–66
  types of, 65–66
Nyquist rate, converters, 87

Object linking and embedding (OLE), 298
One-time programmable (OTP) ROM, 98

On-off keying (OOK), 179
Open-loop system, 150
  accelerometer circuit, 152
  comparison, 151
  See also Control
Open Systems Interconnection
    (OSI), 120, 123
Operation
  rail-to-rail, 76–77
  static vs. dynamic, 57
Optical signal transmission, 182
OSEK protocol, 129–30
Overmolded pad array carrier
    (OMPAC), 234
Overpressure, 56
Oversampling, converters, 87

Packaging
  ball grid array (BGA), 228, 234–35
  ceramic, 232
  design, 229
  DIP, 228
  dual-chip, 233–34
  flip-chip, 237–38
  glob top, 239, 240
  hybrid, 231–35
  microBGA, 228
  for monolithic sensors, 235–39
  PGA, 228
  pinouts, increased, 231
  plastic, 236
  reliability and, 227
  requirements, 228
  semiconductor, 228–31
  SIP, 228
  standards, 245
  surface-mounted, 236–37
  through-hole, 228
  wafer-level, 238–39
Particulate control, surface micromachining,
    30
Partitioning
  possibilities, 9
  system, 250
Parylene deposition, 231
Passivation layer, 230
Performance ADC, 89–90
Peripherals, MCU, 96–97

Recent Titles in the Artech House Sensors Library

Chemical and Biochemical Sensing With Optical Fibers and Waveguides, Gilbert Boisdé and Alan Harmer

Introduction to Microelectromechanical (MEM) Microwave Systems, Hector J. De Los Santos

An Introduction to Microelectromechanical Systems Engineering, Nadim Maluf

Optical Fiber Sensors, Volume III: Components and Subsystems, Brian Culshaw and John Dakin, editors

Optical Fiber Sensors, Volume IV: Applications, Analysis, and Future Trends, Brian Culshaw and John Dakin, editors

Sensor Technology and Devices, Ljubisa Ristic, editor

Smart Structures and Materials, Brian Culshaw

Understanding Smart Sensors, Second Edition, Randy Frank

For further information on these and other Artech House titles, including previously considered out-of-print books now available through our In-Print-Forever® (IPF®) program, contact:

Artech House
685 Canton Street
Norwood, MA 02062
Phone: 781-769-9750
Fax: 781-769-6334
e-mail: artech@artechhouse.com

Artech House
46 Gillingham Street
London SW1V 1AH UK
Phone: +44 (0)20 7596-8750
Fax: +44 (0)20 7630-0166
e-mail: artech-uk@artechhouse.com

Find us on the World Wide Web at:
www.artechhouse.com